MATERIALS AND PROCES

MATERIALS AND PROCESSES

FOR ELECTRICAL TECHNICIANS

BY

L.C.MOTT

A.M.I.E.D.
Senior Lecturer
School of Electrical and Mechanical Engineering
Bordon, Hampshire

CLARENDON PRESS · OXFORD

1974

Oxford University Press, Ely House, London W. 1

GLASGOW NEW YORK TORONTO MELBOURNE WELLINGTON
CAPE TOWN IBADAN NAIROBI DAR ES SALAAM LUSAKA ADDIS ABABA
DELHI BOMBAY CALCUTTA MADRAS KARACHI LAHORE DACCA
KUALA LUMPUR SINGAPORE HONG KONG TOKYO

PAPERBACK ISBN 0 19 859340 6
CASEBOUND ISBN 0 19 859345 7

© OXFORD UNIVERSITY PRESS 1974

Printed in Great Britain by
J. W. Arrowsmith Ltd., Bristol

GENERAL EDITOR'S FOREWORD

This volume has been written specifically to meet the needs of technician students. Although primarily designed for Electrical Technicians, it also contains much of importance for Telecommunications Technicians.

The author has directed his efforts towards presenting the subject as a natural progression rather than separating the material into separate syllabuses and has included material covering the syllabus of both years (T1 & T2) of the C.G.L.I. course.

Students preparing for the T2 examinations need both the T1 and the T2 material. Those preparing for the T1 year will find that the inclusion of T2 material aids their understanding of the subject. Reference should be made to the current syllabus published by the C.G.L.I. if selection of topics is required for teaching or preparation purposes.

Joyden's Wood
1974   Sydney F.Smith

PREFACE

Technicians employed in the electrical industry should be conversant with the large number of different materials used in their work. Also, they should have a knowledge of the methods of production, relative costs, mechanical and physical properties, applications, advantages and disadvantages, and the modern processes available for making the materials into efficient components.

Courses of study have been arranged by examining bodies to give technicians the opportunity of learning both the practical and theoretical aspects of their work.

This book has been written specifically to cover the *Materials and processes* section of *Related Studies* for the *City and Guilds of London Institute's Course 281: Electrical Technicians*. It should also be of interest to students who are following courses of study closely associated with electrical work, such as telecommunications technicians, as well as the interested layman.

The 25 chapters are divided into two sections. The first, comprising 13 chapters, deals progressively with the production, properties, processing and applications of a wide variety of materials including ferrous metals, conductors, insulators, and magnetic materials and semi-conductors. The second deals mainly with workshop processes and includes the use of hand tools, machine tools and soldering and welding equipment. In addition there are chapters dealing with forging, casting, extrusion and other essential forming processes.

The text is illustrated where necessary and numerous tables of properties of materials are included to act as sources of information and also to support the summaries at the end of each chapter.

There are student exercises of examination standard at the end of the book. Their object is to test the reader's knowledge and to cover the homework requirements of the courses. A feature of the book is that students can check their answers by reference to the most relevant section in the text.

Farnborough, 1974   L.C.M.

ACKNOWLEDGEMENTS

The author wishes to acknowledge with gratitude the assistance given by the following organizations:

Aluminium Federation, Portland House, London, S.W.1.

British Iron and Steel Federation, Steel House, Tothill Street, London, S.W.1.

CIBA (ARL) Ltd., Duxford, Cambridge.

Copper Development Association, 55 South Audley Street, London, W.1.

Imperial Chemical Industries Ltd., Plastics Division, Welwyn Garden City, Hertfordshire.

The International Nickel Company (Mond) Ltd., Thames House, Millbank, London, S.W.1.

The City & Guilds of London Institute, 76 Portland Place, London W.1. for permission to reproduce a selection of questions from past examination papers.

The Micanite & Insulators Co. Ltd., Blackhorse Lane, Walthamstow, London, W.17.

Henry Wiggin & Co. Ltd., Wiggin Street, Birmingham, 16.

Thanks are also due to the staff of The Clarendon Press, Oxford.

CONTENTS

I. MATERIALS

1.		PROPERTIES OF MATERIALS	3
	1.1	Elements	3
	1.2	Mixtures and compounds	3
	1.3	Atoms and molecules	4
	1.4	Alloys	4
	1.5	Refractory materials	4
	1.6	Mechanical and physical properties	4
	1.7	Summary	7
2.		PRODUCTION AND PROPERTIES OF CAST IRON	9
	2.1	Iron making	9
	2.2	Types of cast iron	11
	2.3	Summary	12
3.		PRODUCTION AND PROPERTIES OF STEEL	13
	3.1	Raw materials	13
	3.2	The processes used to make steel	13
	3.3	Processing steel	17
	3.4	Plain carbon steel	17
	3.5	Elements in steel	19
	3.6	British Standard En steels	19
	3.7	Summary	20
4.		HEAT TREATMENT OF CARBON STEEL	21
	4.1	Changes in mechanical properties	21
	4.2	Methods of heat treatment	21
	4.3	Summary	24
5.		COPPER AND COPPER ALLOYS	26
	5.1	Raw materials and processes	26
	5.2	Smelting	26
	5.3	Properties of refined copper	28
	5.4	Forms of supply	30
	5.5	Copper alloys	31
	5.6	Summary	33
6.		ALUMINIUM AND ALUMINIUM ALLOYS	34
	6.1	Raw materials	34
	6.2	Refining	34
	6.3	Properties of aluminium	34
	6.4	Alloying elements	37
	6.5	Summary	37
7.		MAGNETIC MATERIALS	38
	7.1	Magnetic properties	38
	7.2	Hard magnetic alloys	38
	7.3	Soft magnetic materials	39
	7.4	Slightly magnetic alloys	40

	7.5	Non-magnetic alloys	41
	7.6	Summary	41
8.	**CONDUCTING MATERIALS**		43
	8.1	Conductors	43
	8.2	Applications	43
	8.3	Copper conductors	44
	8.4	Aluminium conductors	45
	8.5	Electrical contacts	47
	8.6	Circuit breakers	48
	8.7	Carbon brushes	50
	8.8	Relay contacts	50
	8.9	Vacua, gases, and vapours	51
	8.10	Evacuated glass envelopes	52
	8.11	Gas-filled envelopes	53
	8.12	Summary	54
9.	**ELECTRICAL RESISTANCE MATERIALS**		55
	9.1	Classification	55
	9.2	Copper-base alloys	55
	9.3	Iron-base alloys	56
	9.4	Nickel-base alloys	56
	9.5	Summary	57
10.	**BIMETALS AND THERMOCOUPLE MATERIALS**		59
	1. Bimetals		59
	10.1	Sandwich cladding	59
	10.2	Electrical applications	59
	10.3	Silver cladding	60
	10.4	Copper cladding	60
	10.5	Thermostat materials	61
	10.6	The Nilo series of alloys	61
	2. Thermocouples		62
	10.7	Summary	64
11.	**ELECTRICAL INSULATING MATERIALS**		65
	11.1	Paper insulation	65
	11.2	Cambric insulation	66
	11.3	Insulating oil and wax	66
	11.4	Elastomers	67
	11.5	Plastics materials	70
	11.6	Mica	75
	11.7	Vulcanized fibre	77
	11.8	Porcelain insulators	77
	11.9	Glass insulators	78
	11.10	Mineral insulation	78
	11.11	Textile coverings	79
	11.12	Insulation failure	79
	11.13	Summary	79
12.	**TRANSFORMERS, RESISTORS AND CAPACITORS**		81
	12.1	The transformer	81
	12.2	Resistors	82
	12.3	Capacitors	84
	12.4	Summary	87

13.   CABLES, LINES AND SUPPORTS   88
      13.1   Underground electrical cables   88
      13.2   Underground telephone cables   88
      13.3   Electrical transmission lines   89
      13.4   Telecommunications lines   90
      13.5   Line supports   90
      13.6   Summary   91

II. PROCESSES

14.   FOUNDRY PROCESSES   95
      14.1   Metal casting   95
      14.2   Summary   100

15.   FORGING   101
      15.1   Forging principles   101
      15.2   Summary   104

16.   ROLLING   105
      16.1   Hot rolling   105
      16.2   Cold rolling   106
      16.3   Annealing   107
      16.4   Summary   107

17.   DRAWING AND EXTRUSION   109
      17.1   Wire drawing   109
      17.2   Conduit   110
      17.3   Extrusion   110
      17.4   Summary   113

18.   PRESSING AND PRESS TOOLS   114
      18.1   Processes   114
      18.2   Materials   115
      18.3   Summary   115

19.   SOLDERS AND SOLDERING   116
      19.1   Lead   116
      19.2   Tin   116
      19.3   Soft solders   116
      19.4   Fluxes   117
      19.5   Soldering   118
      19.6   Plumbing   119
      19.7   Summary   121

20.   BRAZING AND WELDING   122
      20.1   Brazing   122
      20.2   The brazing process   123
      20.3   Welding   123
      20.4   Summary   126

21.   BENCH WORK   127
      21.1   The basic processes   127
      21.2   Measurement   127
      21.3   Marking out   132
      21.4   The hacksaw   135
      21.5   Filing   136
      21.6   Hand scrapers   138

		21.7	Cold chisels	139
		21.8	Summary	139
22.	THE DRILLING MACHINE			141
		22.1	Cutting tools	141
		22.2	Summary	143
23.	TAPS AND DIES			144
		23.1	Screw cutting	144
		23.2	Summary	146
24.	THE LATHE			147
		24.1	General description	147
		24.2	Particular features	147
		24.3	Work holding	149
		24.4	Turning	150
		24.5	Summary	151
25.	CUTTING TOOLS AND LUBRICANTS			152
		25.1	Tool geometry	152
		25.2	Cutting tool materials	156
		25.3	Coolants and lubricants	158
		25.4	Summary	159

EXERCISES   160

INDEX   167

SECTION I
MATERIALS

1. PROPERTIES OF MATERIALS

1.1 Elements

All materials used in engineering production have complex structures consisting of one or more *elements*. As there are over 90 naturally occurring elements, some of which have complicated names, a universal code of symbols is used to describe them. Some elements and their symbols are listed in Table 1.1.

TABLE 1.1
A selection of elements and their symbols

Element	Symbol	Element	Symbol
Aluminium	Al	Molybdenum	Mo
Antimony	Sb	Nickel	Ni
Argon	Ar	Nitrogen	N
Arsenic	As	Osmium	Os
Beryllium	Be	Oxygen	O
Bismuth	Bi	Palladium	Pd
Cadmium	Cd	Phosphorus	P
Calcium	Ca	Platinum	Pt
Carbon	C	Rhodium	Rh
Chlorine	Cl	Ruthenium	Ru
Chromium	Cr	Silicon	Si
Cobalt	Co	Silver	Ag
Copper	Cu	Sodium	Na
Gold	Au	Sulphur	S
Hydrogen	H	Tellurium	Te
Iridium	Ir	Tin	Sn
Iron	Fe	Titanium	Ti
Lead	Pb	Tungsten	W
Magnesium	Mg	Uranium	U
Manganese	Mn	Vanadium	V
Mercury	Hg	Zinc	Zn

1.2 Mixtures and compounds

Elements which can be intimately mixed without chemically reacting are said to form a *mixture*. A mixture has the properties of the separate elements that form it, these properties being directly related to the quantities of each element contained in the mixture. In such a process no new substance is evolved. The air we breathe, for example, is a mixture consisting almost entirely of oxygen and nitrogen, and the properties of air depend upon the proportions of these gases present in the mixture.

However, both of these gases can be combined chemically with each other, and with other gases, to form substances which are totally different from air. One such special combination of oxygen and nitrogen forms a gas called nitrous oxide, an anaesthetic having properties entirely different from either oxygen, nitrogen, or air. This type of chemical substance is called a *compound*.

1.3 Atoms and molecules

All substances are composed of *atoms* or *molecules* and each molecule is made up of atoms. An element is composed of a single kind of atom, and this condition does not alter even if the state of the element changes from solid to liquid and then to a gas.

Elements unite with one another to form compounds and, as a result, a compound has molecules made up of two or more different types of atoms.

1.4 Alloys

Alloys are metallic substances formed as a result of the intimate blending of two or more different elements. The properties of alloys differ greatly from those of their constituent elements taken separately.

In general, alloys consist of a parent element (metallic), to which is added a smaller amount of another element which can be either a metal or non-metal. For example, carbon steel is basically an alloy composed of a large quantity of iron and a small quantity of carbon. The non-metallic element carbon is added to the iron while it is molten, where it dissolves in much the same way as sugar dissolves in hot tea.

When the metal cools to room temperature, the iron retains the carbon in *solid solution* and the resulting metal alloy has properties that are totally different from either iron or carbon alone. See Fig. 1.1 which shows some common ferrous metals (used to make steel), some non-ferrous metals, and the alloys formed by them.

1.5 Refractory materials

Refractory materials are, generally, non-metallic minerals used in the construction and lining of metal-processing furnaces and ladles. These materials are also used for the insulation of high-temperature electrical equipment and in the manufacture of bricks for electrical storage-heaters. The most widely used minerals are fireclay, quartzite, dolomite, and magnesite.

Refractories must be capable of withstanding high temperatures, and rapid changes in temperature. They must also resist abrasion, erosion, and impact.

1.6 Mechanical and physical properties

The materials used in all forms of engineering have characteristics which make them suitable for their particular job. These characteristics may be divided into two groups; the *mechanical properties* and the *physical properties* (see Table 1.2). Mechanical properties refer to the behaviour of the materials when subjected to forces, pressure, impact, indentation, etc., and the physical properties refer to characteristics such as thermal and electrical conductivity, magnetic properties, and relative density.

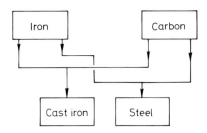

Basic ferrous metals

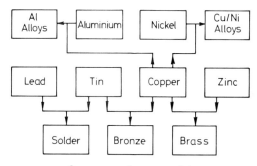

Some non-ferrous metals

Fig. 1.1   Some common ferrous and non-ferrous metals and their alloys

TABLE 1.2
Typical properties of a small selection of materials

Material	Chemical symbol	Tensile strength (N/mm^2)	Percentage electrical conductivity (Silver = 100)	Percentage thermal conductivity (Silver = 100)	Relative density	Melting temperature ($^\circ$C)
Aluminium	Al	100	57	35	2.7	660
Brass	–	300	25	24	8·4	–
Cast iron	–	250	15	13	7·9	–
Copper	Cu	155	94	92	8·9	1083
Gold	Au	75	67	70	19·3	1063
Lead	Pb	15	7	9	11·4	327
Mercury	Hg	–	1·5	5	13·6	−38·5
Nickel	Ni	370	20	14	8·9	1455
Platinum	Pt	325	14	38	21·4	1774
Tungsten	W	4000	34	38	19.2	3375

In order to be of value to the engineering industry and its customers, a material must be capable of being fashioned into shape by machining and other forming processes, without loss of strength, and it is required to give long and efficient service at the lowest possible cost.

Brittleness is that quality of a material which causes it to fracture suddenly when subjected to a shock loading. Sheet glass is a brittle material because it will shatter immediately it is struck a sharp blow.

Toughness is the opposite of brittleness and is the property of a material which describes its ability to resist impact and shock loads, bending, and twisting.

Ductility is the property of a material that enables it to be drawn out in length without breaking. A material that can be drawn into wire form is said to be ductile.

Elasticity is the property of a material that enables it to return to its original shape and size after it has been deformed. A piece of elastic gets its name from its elastic properties. Applying a force which tends to pull the ends apart can stretch it a considerable distance and yet, as the force is removed, the elastic will revert to its original size and shape.

Plasticity is the opposite of elasticity and is the property of a material that enables it to retain any shape imposed by a force after that force is removed. Metals which have this characteristic are used to make medals and coins because they will retain the images and lettering machined into the dies.

Malleability is a property enabling materials to be beaten and mechanically formed into shape without cracking. Malleability should not be confused with ductility. The qualities are similar when considering a material suitable for forming into sheet or strip, but not necessarily when considering drawing into wire form. For example, lead is a malleable material which can be beaten or rolled into thin sheets but it is not suitable for making wire. Malleable materials are those used for forging, stamping, pressing, etc.

Hardness is the property of a material that enables it to resist indentation, scratching, and wear (see Table 1.3). A hard material generally possesses little ductility and is often brittle. Materials such as glass and commercial cast iron are hard and brittle.

Fusibility is the property of a material that enables it to melt when heated. Most metals are fusible, some plastics and refractory materials are not. Tungsten is an example of a metal which is not fusible. This is one reason why it is used as the light source in electric lamps. It can be heated in a furnace until its surfaces become soft and tacky, but it cannot be melted on a commercial scale. Some plastics materials, notably the material loosely called 'Bakelite', will not melt. High temperatures will cause some of the constituents to burn but the plastics material itself blisters and chars.

Conductivity as applied to solids is the ability of a material to conduct heat or an electric current. Both the thermal and electrical conductivity of a material are increased when there are freely moving *electrons* in its structure. The electrons, which

TABLE 1.3
Relative hardness and cost of a selection of materials

Material	Relative hardness (talc = 1, diamond = 10)	Relative cost (p/kg) (1973)
Aluminium	2·5	26
Antimony	3	110
Brass	4	50
Carbon steel	5	4
Cast iron	6	3
Chromium	9	82
Copper	3	60
Iron (soft)	3	3
Lead	1·5	13
Manganese	5	56
Mica	3	12
Phosphor bronze	4	75
Silver	5	3500
Tin	1·5	146
Tungsten	9	560
Zinc	2·5	12

are negatively charged particles, surround the positively charged nucleus of each atom of the structure.

Because the atoms of metallic substances have electrons which are free to move through the entire solid, they are good conductors. Most non-metallic solid substances have structures in which the electrons are closely bound to individual atoms and are not free-moving; as a consequence they are poor conductors.

1.7 Summary

Metals or non-metals made up of one type of atom are called elements. A mixture of elements has the properties of the separate elements that form it and no new substance is evolved. A compound is formed from the chemical combination of two or more elements. The properties of a compound are different from those of the separate elements forming it. All substances are composed of atoms either uncombined, or combined with other atoms to form molecules. An element has molecules containing exactly similar atoms, whilst a compound has molecules composed of two or more different atoms.

Alloys are made by dissolving one or more elements in another (metallic) element while molten, the alloying elements being retained as a solid solution on cooling.

Refractory materials are those capable of withstanding very high temperatures without melting.

Mechanical properties of materials refer to their behaviour when subjected to mechanical forces and pressure. Physical properties of materials refer to their behaviour when subjected to heat, magnetic fields, electric currents, etc., and also to their relative densities and melting temperatures.

2. PRODUCTION AND PROPERTIES OF CAST IRON

2.1 Iron making

Iron ore is found mixed with stone and earthy matter either on, or just below, the surface of the earth. To separate the iron from the unwanted matter, the ore is mixed with large amounts of coke and small amounts of limestone and charged into a *blast furnace* of the type shown in Fig. 2.1.

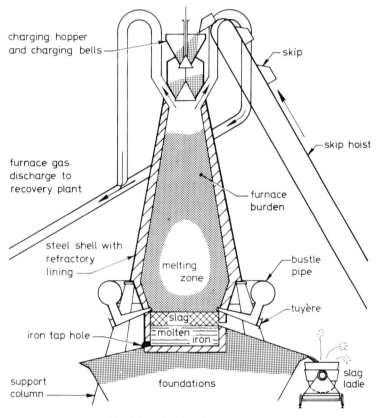

Fig. 2.1   The blast furnace

The charge of ore, coke, and limestone is carried to the top of the furnace in special containers called *skips*, where it is tipped into a specially designed hopper fitted to the top of the furnace. This is arranged so that when the charge drops into the furnace none of the gases escapes into the atmosphere.

The blast furnace is a huge steel vessel lined with special refractory bricks to retain the heat. Furnaces can be as high as 30m and with a diameter at the base of 10m.

Close to the base or furnace hearth is a large-diameter pipe which encircles the furnace. This pipe, called the *bustle*, carries hot air under pressure to a number of smaller pipes which enter the furnace through holes called *tuyères*. The blast of air acts as a huge bellows and the oxygen it contains causes the coke to burn with an intense heat that melts the iron ore. The limestone assists in the melting process and also reacts chemically with the earthy matter to form a waste product called *slag*, which separates from the iron.

As the iron and the slag melt they filter through the burning coke and drop to the furnace hearth where they collect. The slag is lighter than the molten iron and so it floats on top of the pool of liquid metal. The temperature in the furnace at its hottest point is about 1800°C, and that required to melt the iron is about 1500°C.

When sufficient molten iron has collected in the hearth the clay plugs filling the tapping holes are drilled out, the furnace is emptied of its molten iron and slag and the holes are re-plugged; the process is continuous.

Molten iron flowing from a blast furnace can be treated in two ways. When produced for immediate steel making it is run directly from the blast furnace into huge ladles. It is then transferred to another furnace for processing into steel. When the iron is intended for the production of cast iron it is run into open channels or moulds made of sand, where it cools into ingots of *pig iron* or *pig*.

This is the traditional method of pouring the metal to allow it to solidify. In a modern smelting plant a pig-iron moulding machine is used. This is composed of an endless and slow-moving conveyor attached to which is a series of open-topped moulds. The molten iron is poured into the moulds and water sprays help to solidify the metal as the conveyor moves slowly up an incline. At the top of the incline the moulds tip over, releasing the solid block of iron into a waiting railway wagon. The iron is then transported to steel works or iron works where it will be further processed.

Cast iron production in Great Britain is limited to about one tenth of the output from the blast furnaces, the other nine-tenths are processed into steel.

The pig iron is first melted in another type of blast furnace, called a *cupola*, a diagram of which is shown in Fig. 2.2. This furnace is much smaller than the blast furnace used to process iron ore, and it is here that the quality of the cast-iron products will be decided. When the furnace is tapped, the molten cast iron flows into ladles from which the metal is poured directly into moulds.

Other elements in cast iron come from two sources; those in association with iron while it is in the earth, and those absorbed during the time spent in the furnace. These elements are phosphorus, silicon, sulphur, manganese, and carbon. The amounts of these elements present in the molten cast iron flowing from the cupola will, to a great extent, determine the quality of the casting. The most important of these elements is carbon, which varies in amount between 2·5 per cent and 4·5 per cent depending upon the type of cast iron required. It contributes to the hardness and brittleness of the metal and as the amount used increases, the melting temperature of the cast iron decreases.

Although the other elements present play a part in giving cast iron its particular qualities, they are generally considered to be impurities and are retained only in very small amounts.

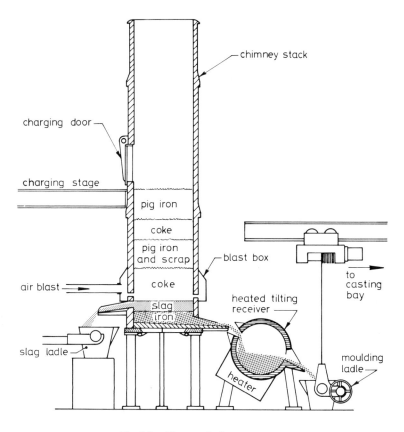

Fig. 2.2   The cupola furnace

The form of supply of cast iron is only as castings; it cannot be processed into sheet, rod, bar, or sections, except in cast form.

2.2 Types of cast iron (see Table 2.1)

The form of the carbon in cast iron will largely determine its characteristics and its uses. In ordinary commercial cast iron the carbon exists in two forms, as iron carbide and free graphite flakes.

White cast iron is produced when castings are very quickly cooled and the carbon has no time to separate from the iron. The metal is hard and brittle and machining is difficult. When the iron is broken the fracture displays a silvery appearance and this gives rise to the name white cast iron.

Grey cast iron is produced when castings are slowly cooled in the mould. This allows time for the carbon to settle out in the form of free flakes of graphite. The resulting iron is softer and more easily machined. When the iron is broken, fracture takes place where the graphite flakes are located, displaying a dark grey colour.

TABLE 2.1

Typical properties and applications of a small selection of cast irons

Type of cast iron	Approximate carbon content (%)	Approximate specific electrical resistance ($\mu\Omega$ m)	Approximate tensile strength (N/mm^2)	Typical applications
Grey cast iron	3	0·75	300	Frames for electric motors and generators. Baseplates, brackets, and bearing blocks
Malleable cast iron	3	0·30	355	Switch boxes, fuse boxes, etc., conduit fittings
Nodular cast iron	3	0·50	695	Transformer parts, cable boxes, switch boxes and covers, busbar clamps, electrode holders, circuit breaker covers, etc.

Malleable cast iron is made by heating white cast iron in a furnace over a consider-able period of time. This heat treatment creates the conditions for the formation of small clusters of carbon particles, finely dispersed throughout the casting. Cast irons of this type are used for components which would otherwise have to be made from a more expensive material, or manufactured using costly processes.

Junction boxes, conduit fittings and switch boxes are made from malleable cast iron. The material is tough and relatively cheap, but the disadvantages are that it can be used only for components having comparatively thin sections, and the cost in time and money of the long periods that the castings have to spend in the furnace in order to change the cast iron into the malleable form.

Although called malleable cast iron, this does not mean that the metal can be beaten or rolled into shape; it is only supplied in cast form. However, it is tougher than cast iron and cheaper than steel.

Nodular cast iron is produced by introducing a small percentage of magnesium into the molten metal prior to casting. The graphite forms into small nodules which are dispersed fairly evenly throughout the mass of iron. Magnesium burns readily, however, and to ensure that sufficient of this element is retained in the molten cast iron, an alloy of magnesium and nickel is used. The nickel also improves the characteristics of the iron, which can be used for a number of applications traditionally reserved for steel.

2.3 Summary

Iron is produced in a blast furnace by melting together iron ore and limestone, the furnace fuel being coke. The iron is either made into pig iron or processed directly into steel. Cast iron is produced by melting pig iron in a cupola furnace. Malleable cast iron, used in the production of electrical conduit fittings, is produced by the prolonged heating of white cast iron. Iron, its alloys, and the metals used to make its alloys are called *ferrous metals*.

3. PRODUCTION AND PROPERTIES OF STEEL

3.1 Raw materials

Steel is essentially an alloy of iron and carbon and there are two main types of iron used in steelmaking. These are *basic iron*, which is produced from ores having a high phorphorus and low silicon content, and *haematite* or *acid iron*, which has a low phosphorus content. More basic iron is produced than acid iron owing to the greater abundance of basic iron ore.

Iron is processed into steel by refining in a furnace. This either eliminates or considerably reduces the carbon and other elements considered harmful to steel. To assist in this process quantities of limestone are used together with specially prepared furnace linings.

3.2 The processes used to make steel

The processes used to make steel are: the *converter* processes, so-called because they convert iron into steel, the *open-hearth* process, which is described in § 3.2.2, and the *electric furnace* process, which is described in § 3.2.3.

The open-hearth furnaces account for at least 75 per cent of steel production in Great Britain, followed by about 15 per cent for the various converter processes, and the remaining 10 per cent is produced by electric furnaces and other small quantity processes.

3.2.1 The Bessemer converter

The Bessemer converter is a large pear-shaped vessel, somewhat like a huge concrete mixer, a diagram of which is shown in Fig. 3.1. It is made from heavy steel plates riveted together and lined with refractory materials.

The lining mixture, which differs for the production of acid and basic steel, is firmly attached to the inside surfaces of the vessel and baked using a coke fire.

The converter is mounted on two side projections called trunnions which enable it to be tilted for the charging, sampling, and pouring operations. An air passage is provided in one of the trunnions, and this leads to an air box attached to the base of the converter. Holes are made in the bottom of the vessel and the refractory lining to act as tuyères and allow a blast of high-pressure air to pass upwards through the molten charge.

Once the coke fire is at the correct temperature, steelmaking may begin. The converter is tilted to one side and charged with burnt limestone, molten pig iron and scrap steel. The limestone helps to remove the unwanted matter into the slag.

When charged, the vessel is turned to an upright position and the air blast is turned on, making the molten material 'burn' very fiercely. This stage is called the *blow* and a huge flame spurts from the mouth of the converter and passes through a chimney and out to the atmosphere. All the violent changes taking place inside the converter are brought about by chemical reaction and the air blast, which helps generate the heat required to burn out or *oxidize* the unwanted elements.

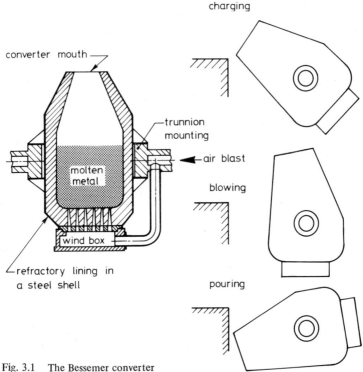

Fig. 3.1   The Bessemer converter

Carbon monoxide gas is evolved within the converter during the blow and as this is released it causes the molten metal to *boil*. After about 20 min the boiling action slowly subsides as the carbon in the iron becomes spent, and the steelmaker watches the flame at the converter mouth die away and change to brown smoke as the phosphorus is also eliminated. This latter stage in the process, which takes only a few minutes, is called the *afterblow*.

Shortly after this the air blast is shut off, the converter is tilted and a sample of the molten metal is taken. This is poured into a mould and cooled. The sample is tested to see if further blows of the converter are required.

The Bessemer process of converting iron into steel takes about 30 min, after which the converter is tilted to pour the slag and the steel into separate ladles.

Carbon in the steel is eliminated during the blow and the correct amount required for the grade of steel being produced can be added to the molten metal in the converter prior to pouring. Any other elements that may be required can be added in the ladle after pouring.

The *Bessemer processes* do not produce the highest quality metal. General-purpose steels only, known as *mild steel* and *medium carbon steel*, are produced in this way. There are two different Bessemer processes, called the *basic* and the *acid*. They derive their names from the type of iron used, the furnace lining material, and the slag produced.

Basic Bessemer steel is made from iron rich in phosphorus. The converter requires a lining of crushed dolomite and tar which is rammed into place against the inside surfaces of the vessel. The dolomite, which is a naturally occurring rock, counteracts the harmful effects of the basic slag produced during steelmaking which would otherwise attack the vessel lining.

Acid Bessemer steel is made from iron having a low phosphorus content. The converter uses a silicon brick lining and an acid slag is formed.

3.2.2 Open-hearth furnace

The open-hearth method of converting iron into steel was started in England after the Bessemer process was established. Its main advance was in the use of waste heat from the furnace to provide a steady flow of pre-heated air to the furnace fuel in order to generate higher temperatures in the hearth.

In Great Britain the majority of steel is produced in open-hearth furnaces. The pig iron and scrap are melted in an atmosphere of extremely hot gases which pass over the shallow open hearth. In addition to melting the charge, the furnace gases also remove the carbon and other impurities. Because careful sampling and control can be exercised, a good quality steel is produced.

Fig. 3.2 is a diagram of an open-hearth furnace. The furnace consists of a large shallow bath having a solid back wall and an arched roof. The front of the furnace is fitted with doors through which the hearth is charged with iron, limestone and scrap metal. The refractory brickwork required to retain the heat is supported by steelwork and the furnace can be either in a permanently fixed position or it can be of the tilting variety.

Referring to Fig. 3.2, the method of operation is as follows. The air valve is set to allow air to enter the left-hand brickwork chamber only, the walls of which have been heated during a previous cycle of operations. The air passing upwards into the furnace is heated through contact with the chequered brickwork and joins the flames from the fuel burners. Tremendously hot gases pass across the charge in the hearth and down into the right-hand brickwork chamber, which absorbs the heat from the burnt gases. When the heat stored in the left-hand chamber is used up, the burners are turned off and the air inlet valve is reversed; the right side-burners are lighted, and the air now passes into the furnace via the very hot right-hand chamber. The direction of the air flow is reversed in this way at regular intervals, and the burners to the left and right are used alternately to ensure maximum steel production with minimum loss of heat.

Open-hearth steels are made from the basic or acid irons previously described in § 3.1.

In the basic process a certain amount of limestone is added to the furnace charge in order to provide a slag which will carry away unwanted matter and absorb the harmful elements. By the time the furnace charge has completely melted, a great deal of the carbon and silicon have been removed by oxidation. This is achieved by the oxygen in the flame which burns out these elements as they combine with the oxygen and leave the furnace as a gas.

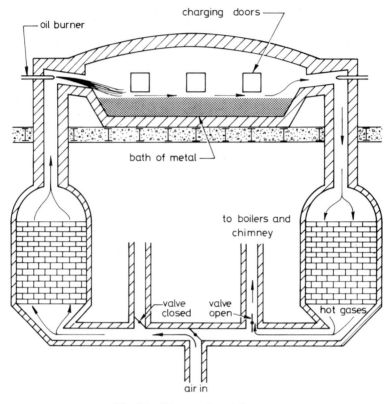

Fig. 3.2   The open-hearth furnace

To remove the remaining carbon, silicon, phosphorus and sulphur, oxidizing agents such as iron ore and *mill scale* are added to the molten charge. Mill scale is the flakes of steel ingot while it is passing through the rollers of a rolling mill. During this latter oxidizing process, known as the refining period, samples are taken of the metal and the slag to establish the quality of steel and to enable the steel-maker to judge the correct time to tap the furnace.

The additional elements which may be required to produce a particular grade of steel, such as an alloy steel, are introduced after the metal has been poured from the furnace into a ladle.

A typical open-hearth furnace can produce about 100 000 kg of steel in 12 hours.

3.2.3 Other processes

The *duplex process* is a combination of the Bessemer and open-hearth processes, often used to produce high quality steels. The converter removes most of the impurities including the carbon from the charge, which is then transferred to the open-hearth furnace to further refine the steel.

The *electric furnace* is used either to refine a steel transferred from the Bessemer converter, or to melt a charge of selected iron and scrap to produce a steel of high quality. This type of furnace allows very accurate control of the atmosphere and temperature, but it is expensive to operate. For this reason its use is often limited to the production of steel for continuous casting, alloy steels and special alloys.

The heat required for the melt is provided by an electric arc struck between electrodes fitted above the hearth, and the charge.

Crucible steel is of high quality because it is made in a non-oxidizing atmosphere and is relatively free from harmful impurities. The high grade of iron and other elements used are melted in a covered crucible and subsequently poured into ingot moulds. At this stage the steel is left for a few hours to cool and settle very slowly, so that any impurities gradually rise and mix with the slag.

Crucible steel, which is very expensive, is used for engineering tools and surgical instruments.

3.3 Processing steel

When the composition of molten steel is acceptable to the steelmakers, it is poured or *teemed* very carefully into huge tapered sleeves made of iron, which stand on heavy base-plates. These sleeves are moulds which form ingots of steel. Teeming is a highly skilled operation because the liquid steel must emerge from the bottom of the ladle in a thin pencil-shaped stream. Splashes which would spoil the ingot must be kept to a minimum.

Ingots are processed into shapes and sizes which can either be stored for future use or manufactured directly into things such as sheets for motor car bodies or tinned plate. The ingot is the starting point of almost all steel articles and is generally of a tapered rectangular section with a short length of an unusable mixture of metal and slag at the top. This is called the *feeder head* and is where the slag collects during cooling. The feeder head is cut off before the ingot is rolled or forged.

A flow chart is shown in Fig. 3.3 which gives some idea of the various sections and associated products made from a steel ingot.

Ingots intended for the rolling mills are reduced in size by heavy rollers which make them into either *blooms* or *slabs*, depending upon the other processes through which they must go before taking their final shapes. Blooms are made into sheet bars and then sheets, large rounds, small slabs, flats and billets and steel sections. Slabs are processed into plates, sheet and strip.

Cast steel is not a special form of steel; it is steel produced in cast shapes which are too difficult to be carried out in the forge. Such castings are often produced in the steel works direct from the ladle of molten metal. Machine tool and locomotive frames and castings for ships, heavy goods vehicles, and large electrical generator bodies are some of the items produced in this way.

3.4 Plain carbon steels

Steel is basically an alloy of iron and carbon, and plain carbon steels are classified according to the percentage of carbon contained in the metal.

Low-carbon steel contains carbon in amounts up to 0·25 per cent. This grade of steel is used for a wide variety of general engineering applications. It is made into

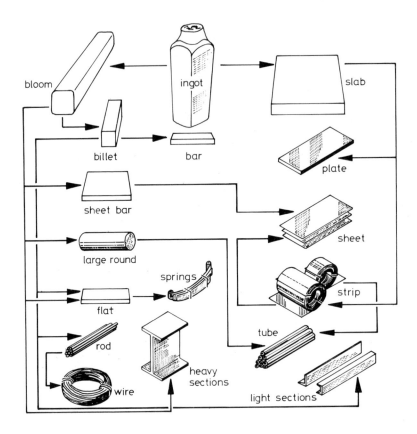

Fig. 3.3   Steel products from an ingot

plate, sheet, strip, bar and rod, and also rolled sections such as angles, channels and joists. It is machined into bolts, nuts, washers and rivets.

Low-carbon steel is often called *mild steel*. It is easily machined and welded but will not respond to heat treatment other than case hardening.

Medium-carbon steel contains carbon in amounts between 0·25 per cent and 0·65 per cent. Steel in this category is generally the shock-resisting steel and is widely used for forging crankshafts, machine-tool parts, motor-vehicle axles, connecting rods, bolts, gears and vehicle springs. This type of steel will respond to heat treatment becoming stronger and harder.

High-carbon steel has a carbon content between 0·65 per cent and 1·5 per cent. Steel in this category is tool steel which responds readily to heat treatment and will become very hard but brittle. *Silver steel* is high-carbon steel supplied in bar form with a diameter held to very close limits of accuracy. There is no silver in this steel, the name is derived from its bright appearance.

3.5 Elements in steel

Because steel is made from iron, the elements in iron are present in steel. All are reduced in amount by the conversion process and in some cases particular elements are added to the melt to provide certain characteristics in the final metal.

Carbon is the most important single element in steel. It helps to control the hardness and strength of the metal and affects the choice of the methods of working the steel into shapes.

Silicon in amounts below 0·2 per cent has little effect on low-carbon steels. Above this amount ductility is reduced and the strength and corrosion resistance are improved. For many electrical components steel is required to have certain magnetic properties. Silicon in amounts up to 5 per cent gives a steel these properties and also reduces considerably the breakdown of the steel owing to the phenomenon known as *ageing*, a process of gradual deterioration of the structure of the metal. Transformer laminations are made of steel containing at least 4 per cent silicon, to eliminate ageing and the consequent loss of strength and electrical efficiency. Silicon steels are used to make armatures and other similar small parts in motors, generators and dynamos, and also vehicle springs and engine valves.

Manganese helps to purify steel by inducing sulphur to enter the slag by combining with it—the compound then enters the slag. It also improves the welding qualities of steel. Certain percentages of manganese added to a carbon steel will allow sections to be heat treated in such a way that an even hardness can be achieved for a considerable depth. Also, higher temperatures can be used safely during forging, giving increased ductility and improved flow characteristics. A correct balance between the carbon and manganese contents will impart free-cutting qualities, giving a short chip and good finish.

Carbon-manganese steels are used for forgings, bolts, nuts and studs, small lightly-stressed components, and parts which are to be welded.

Phosphorus is introduced into steel from the iron ore and is considered to be an impurity. It has the tendency to make steels brittle even in very small quantities when the steel has a high carbon content. Amounts up to 0·15 per cent can be tolerated in a low-carbon steel because it will slightly improve the strength and corrosion resistance.

Sulphur is an impurity derived from the blast-furnace fuel. It causes brittleness and welding difficulties. Manganese helps to control the small amounts left in steel after it leaves the converter. Free-cutting steels retain about 0·2 per cent sulphur because this small quantity helps to provide a smooth finish and short chip.

3.6 British standard *En* steels

Because it is possible to make an infinite variety of steels, the British Standards Institution, in co-operation with the iron and steel industry, produced Specification *BS 970:1955*, dealing with the rationalization of the numerous specifications then in existence. The steels are classified under En numbers; En being the abbreviation of the term 'emergency number' and dating from the emergency of World War II. Under the appropriate En number will be found the composition of the steel, its mechanical properties, its reaction to heat treatment, welding and machining, and its suitability for use under certain in-service conditions.

A small number of abbreviated steel specifications is given in Table 3.1.

TABLE 3.1

Composition, strength, and uses of a small selection of carbon steels

En No.	Typical composition (%)					Tensile strength (N/mm^2)	Type of steel
	C	Si	Mn	S	P		
1A	0·15	0·1	1·2	0·3	0·05	450	Free-cutting, used for low-duty bolts, nuts, studs, etc.
1B	0·15	0·1	1·4	0·5	0·05	400	Same as 1A
2	0·20	–	0·8	0·05	0·05	300	General purpose cold-forming steel
3	0·25	0·3	1·0	0·05	0·05	385	'20' carbon. Lightly stressed bolts, brackets, etc.
8	0·45	0·3	1·0	0·05	0·05	585	'40' carbon. machine parts, crankshafts. Good wear resistance
9	0·60	0·3	0·8	0·05	0·05	775	'55' carbon. Cylinders, gears, machine tool parts, etc.

3.7 Summary

Steel is essentially an alloy of iron and carbon. It is made in converters, open-hearth furnaces, crucibles, or electric furnaces. Sometimes, combinations of these processes are used to produce special steels. Molten steel is poured into a metal mould to form an ingot. Ingots are processed in rolling mills to reduce them to commercially usable shapes and sizes.

The elements carbon, silicon, manganese, phosphorus and sulphur are all present in steel and are carefully controlled to ensure good quality products. The British Standard for steels is *BS 970*.

4. HEAT TREATMENT OF CARBON STEEL

4.1 Changes in mechanical properties

Heat treatment is carried out on steel for two main reasons; to render the metal suitable for mechanical working and shaping, or to obtain a condition in a finished part that will ensure its correct functioning in service.

A steel ingot requires to be heated so that it can be rolled and squeezed into the more manageable shapes that make up the forms of supply of steel products. Components made from this steel might at some future time undergo a heating and quenching cycle in order to obtain a particular condition within the steel. In both cases the mechanical properties of the steel are being changed for specific reasons.

To obtain the required mechanical properties by heat treatment, a plain carbon steel must contain the correct quantity of other elements, especially carbon. The temperature in the furnace will require careful control, and the method and rate of cooling of the steel when it leaves the furnace must also be controlled.

The arrangement of matter inside a piece of metal is called its *structure*, and it is this structure of tiny crystals or grains, which is altered by heat treatment to give variations in the mechanical properties. All metals are crystalline and prolonged heating of steel coarsens this grain structure, making it mechanically weak.

Consider two thin pieces of steel that are identical in chemical composition except for the carbon content. One piece has 0·12 per cent carbon and the other 0·85 per cent. When both have been heated together in a furnace to a cherry-red colour (just below 800 °C) and quenched in water, they display different properties. The steel specimen with 0·12 per cent carbon will be soft and easily filed. The other specimen with the higher carbon content will be hard and resist attempts to file it. The low-carbon steel will be quite tough but would bend before breaking. The high-carbon steel will be brittle and would snap rather than bend.

4.2 Methods of heat treatment

The term *heat treatment* usually means a hardening process to increase both hardness and strength, followed by tempering which removes stresses caused during hardening. There are four principal methods of heat treatment: *annealing, normalizing, hardening,* and *tempering.* The operations described here will be confined to the treatment of carbon steel.

4.2.1 Annealing

Annealing is a softening operation carried out to enable metal to be worked.

Pick up a piece of old hacksaw blade made of steel and bend it gently between the fingers; you will realize that it is tough and springy. Now hold the blade in the jaws of a pair of pliers and place it in a flame, then allow it to cool down gradually in the warmth of the flame. Once it is cold enough to handle, hold the blade between the fingers as before and bend it. You should now find that the metal is annealed and is quite ductile. It will not only bend easily but will remain bent.

There are three principal annealing operations and these are: *process annealing, full annealing,* and *spheroidize annealing.*

Process annealing will soften the metal and relieve internal stresses. This method of annealing is often carried out on low-carbon steel sheet and wire which has been cold worked. During the working processes the metal hardens and its internal structure is in a state of stress.

The steel is heated in a furnace to a temperature between 550 °C and 650 °C which is held for a length of time depending upon the size of the part being treated and the amount of carbon in the steel. It is allowed to cool slowly in the furnace.

Full annealing is carried out to overcome the problem of brittleness, which often occurs in a steel which has been worked while hot. During the working processes the grains composing the structure of the steel tend to enlarge. Full annealing reduces or refines the grain structure and improves the ductility and machinability of the steel.

The temperature and time in the furnace must be carefully controlled and vary considerably from one steel to another, depending upon the carbon content of the steel and the mass of metal being processed, details of which can be obtained from the steel supplier.

Spheroidize annealing is carried out on high-carbon steels so that they can be worked. The annealing temperature is between 650 °C and 700 °C. A change takes place in the structure of the steel which enables it to be machined and drawn out in length while cold (cold drawn).

4.2.2 Normalizing

Normalizing is carried out to improve the mechanical properties of the metal by refining its grain structure. The process also relieves stresses in the steel set up during other heat treatments or by cold working.

The temperature ranges for different steels to be normalized are similar to those observed for full annealing and are held for a period of time which will ensure complete heat penetration of the mass of metal in the furnace.

The main difference between annealing and normalizing lies in the rate and method of cooling. Normalized components are cooled in still air and reach room temperature more quickly than annealed components, which are allowed to cool slowly inside a cooling furnace.

4.2.3 Hardening

Hardening is a process by which steels are made hard by heat treatment so that they will withstand abrasive conditions and resist scratching and indentation. All steels cannot be hardened in the same way, as shown in Fig. 4.1, and consequently they will not display the same mechanical properties, except for the general one of having a hard outer skin.

Consider the piece of hacksaw blade which was annealed (§ 4.2.1). Hold one end of the piece of metal over a flame as before and heat it until it is red-hot. Maintain this condition for about 15 s and then plunge it quickly into a dish of cold water.

When it is cold try to bend it between the fingers. You should find that it is hard and brittle and will snap easily.

4.2.4 Case-hardening

Case-hardening is a process of forming a hard skin on low-carbon steels. These mild steels contain up to 0·25 per cent carbon and, owing to this low carbon content, the structure of the steels will not respond to a full hardening process. However, steels in this class containing a maximum carbon content of 0·18 per cent can be successfully case hardened by increasing the carbon content to about 0·85 per cent for a very short distance into the metal. This treatment will increase the resistance to surface abrasion but any improvement in tensile strength which might occur will be very small.

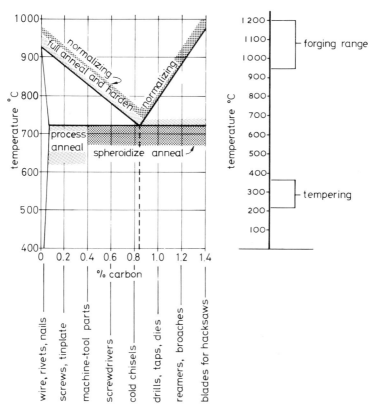

Fig. 4.1   Heat-treatment temperatures for carbon steel

The case-hardening process comprises two principal stages: *carburizing*, when the carbon is induced into the surface of the metal, and *heat treatment*, when the grain structure of the case and the core are modified to produce the required properties in the steel.

Carburizing compounds can be solids, liquids, or gases. The method described here will be confined to the use of solids.

The components to be hardened are placed in a steel box and completely surrounded with a carburizing agent such as charcoal or powdered coke. The lid of the box is sealed using clay, and the packed box is placed in a furnace at a temperature between 900°C and 950°C. The length of time in the furnace will depend on the temperature, the depth of case required, and the grade of steel, but an approximate indication can be obtained by observing that at 900°C, an hour should be allowed for each 0·25 mm of case depth.

During its time in the furnace the steel absorbs the carbon from its surroundings, the carbon being dissolved by the steel. At the higher end of the carburizing temperature scale the carbon is dissolved more quickly and the case is deeper compared with the results at lower temperatures.

Carburized steels can be cooled slowly by allowing the box and contents to cool to room temperature, or quenched by plunging into water or oil, according to the requirements of the finished product. Quenched components are hard whilst those slowly cooled will be soft and capable of being filed.

Core refining is necessary to refine the structure and restore the mechanical properties after the steel has been subjected to a prolonged heating in a furnace during the case-hardening process.

In order to refine the core which is a low carbon steel, the metal is placed in a furnace at a temperature of about 870°C and soaked for a period of time depending upon the carbon content. Immediately upon removal from the furnace the steel is quenched in water or oil.

Case refining is another essential process after case hardening and core refining. The temperature reached in the furnace during core refining coarsens the grain structure of the case which becomes brittle and can flake off.

The steel is reheated to about 780°C and upon reaching this temperature quenching takes place in water. The case will now have the desired degree of hardness and, because the temperature is below that required to alter the structure of the core, this will be unaffected.

4.2.5 Tempering

Tempering is carried out on hardened steel to relieve any stresses due to the severe quenching required to induce hardness and to toughen the metal. The process entails heating in a furnace at temperatures considerably lower than those required for hardening. For example, after the core and case refining procedures detailed above it is often necessary to temper at about 150°C in order to relieve the quenching stresses.

In plain carbon steels the rate of cooling is not important.

Consider again the piece of hacksaw blade in the hardened condition in which we left it in § 4.2.3. It must now be reheated to about 300°C and allowed to cool in the air. The metal should be hard, tough, and springy, in fact restored to its original condition.

4.3 Summary

The softening processes of heating followed by very slow cooling are called annealing. The grain-refining process similar to annealing but with a higher cooling rate is called normalizing.

Heating suitable steel and quenching in water or oil produces hardness and strength. Case-hardening is achieved by inducing carbon into the surface of low-carbon steel; this process being followed by refining heat treatments. Tempering takes place after hardening in order to relieve the stresses imposed by the heat treatment and to toughen the steel.

5. COPPER AND COPPER ALLOYS

5.1 Raw materials and processes

Copper is sometimes found in an almost pure state but this does not account for a great deal of production, the ores being more often discovered as small grains embedded in rock and surrounded by unwanted matter or 'gangue' from which it must be separated. Ores of this type yield only about 4 per cent copper and are found in association with other minerals, notably sulphur, iron, and nickel, some of which are recovered during the processes involved in winning the copper.

Copper has a very strong affinity for sulphur so that generally the ores are sulphides; indeed practically 50 per cent of the world's supply of copper is obtained from an ore which is a sulphide of copper and iron.

5.1.1 Ore dressing

Ore dressing is the name given to the processes involved in preparing the ores for smelting.

In the case of sulphide ores the metal-bearing rock is first crushed into manageable lumps about 125 mm across, then fine-crushed and finally ground to a powder.

The powdered ore next goes to the flotation plant. This consists of a series of tanks through which the ore flows in a continuous stream of water containing special oils which form a froth on the surface. The sulphide ores become coated in the oil and collect in the froth, while the unwanted matter sinks to the bottom of the tanks and is carried away. As the froth rises in the tanks it overflows and carries the copper-bearing material with it. The mixture of copper and other particles is called *concentrate*. The concentrate must have as much water as possible removed before smelting and this is achieved by passing the concentrate through filters where the water is sucked away by a vacuum pump.

Concentrate often contains a vast amount of sulphur and other impurities. To remove these the concentrate is mixed with limestone and silica and *roasted* in multi-hearth roasters. This process prepares the material for the furnace, the limestone and silica being used to promote the formation of a liquid slag.

5.2 Smelting

Almost pure copper ores can be processed in a furnace without pre-treatment because there is no great difficulty in removing the gangue as a slag, but the sulphide ores discussed above require a number of furnace treatments.

The main furnace is called a *reverberatory*. When the roasting process has been found to be unnecessary (when the concentrate is pure) the furnace is charged directly with wet concentrate and quantities of limestone and silica. When roasting has taken place the hot concentrate charged into the furnace is called *calcine*.

5.2.1 The reverberatory furnace

The reverberatory furnace is a large single-chamber structure of refractory bricks strengthened externally by steel sections as shown in Fig. 5.1. The low brick roof is suspended on steel hangers from overhead beams. The interior of the furnace can measure 36 m (120 ft) long, 9 m (30 ft) wide and 4 m (12 ft) high. Running the length of the furnace, along the sides and above the roof, are series of hoppers and feed pipes which allow the concentrate to enter the chamber and form mounds which build up against the furnace walls.

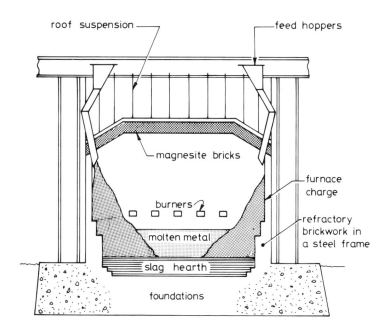

Fig. 5.1　A reverberatory furnace

At one end of the furnace are a number of burners through which the fuel and the air supply enter together. The burners shoot an extremely hot flame the length of the furnace above the charge, the temperature inside the furnace reaching about 2800°C. The radiation from the walls of the furnace melts the concentrate which forms into a mixture of copper sulphide and iron sulphide, called *matte*, and a covering layer of slag. The slag and the matte are run off through separate tap holes.

5.2.2 Converting

Converting is carried out in a manner similar to that used in steel production.

The matte is charged, while still molten, into a converter in order to remove the iron and the sulphur. The converters used are large horizontally mounted cylinders made of steel and lined with refractory bricks. They are mounted on rollers at the ends so they can be tilted for charging, blowing and pouring. Tuyères are provided, connected by pipes to an air pressure supply. Cold air is blown through the molten metal but the chemical reactions inside the converter generate the heat required to purify the matte. The iron and sulphur are oxidized, the iron is removed in the slag and the sulphur is blown out of the vessel as sulphur dioxide gas.

A blow in the converter takes about 3 hours after which the molten copper is poured into ladles and transferred to a casting furnace.

5.2.3 Blister copper

Blister copper is produced in a casting furnace which is heated to maintain the metal in its molten state and also acts as a reservoir. As the copper leaves the casting furnace it falls into a shallow trough called a spoon. Moulds pass beneath the lip of the spoon and the copper overflows into them, where it solidifies. Sulphur gas is evolved within the solidifying copper and this forms bubbles or blisters on the surface of the cast metal cakes giving rise to the name *blister copper*.

At this stage the copper may be between 98 per cent and 99·5 per cent pure; it is also brittle and must be further refined because it is not in a fit condition to be fabricated.

5.2.4 Fire refining

Fire refining of blister copper takes place in a reverberatory refining furnace where the remaining sulphur is driven off as sulphur dioxide gas and the other impurities are removed in a slag. These processes of oxidation are assisted by blowing air through the molten metal which becomes saturated with oxygen. Since oxygen is harmful when present in copper in large quantities, the next stage, called *poling*, is designed to remove the major part by reducing the copper oxide to copper. In this process large quantities of 'green' hardwood tree trunks and thick branches are introduced deep into the molten copper and the surface of the charge is covered with a layer of coke having a low sulphur content.

As the reactions of this process continue, samples of the molten copper are taken and tested by observing the shape of the surface of the solid metal. When this is flat or slightly convex the oxygen content is down to about 0·05 per cent and the copper can be cast either into wire bars or anodes, which are flat slabs or thick plates.

5.2.5 Electrolytic refining

Electrolytic refining enables any precious minerals in association with the copper to be recovered, in addition to its main function of producing high-purity copper.

The refining is carried out in large lead-lined wooden or concrete tanks containing an electrolyte which consists of a heated mixture of dilute sulphuric acid and copper sulphate. Blister copper anodes are suspended in the electrolyte a short distance above the bottom of the tanks and have between them thin sheets of pure copper called *starting sheets* which act as the cathodes. Direct current is supplied to each tank, the anodes being connected to the positive terminals and the cathodes to the negative terminals. When the current is switched on the anodes slowly dissolve and deposit pure copper on the starting sheets. The impurities fall to the bottom of the tanks where they can be collected and separated into recoverable material and sludge. The copper refined in this way can be 99·99 per cent pure. Fig. 5.2 shows a simplified flow diagram of the production of copper.

5.3 Properties of refined copper

The refining methods described above produce what is called *tough pitch copper* of high purity.

Copper is the only red-coloured metallic element in existence. It is soft, extremely malleable, and ductile, making it easy to work, but it does harden when cold worked. It is extruded, rolled, and drawn into varying sections and can be welded, brazed, soldered and tinned. It is a heavy metal having a density of nearly 9×10^3 kg/m^3, a melting point of 1083°C, and a low electrical resistivity of $0.017\,\mu\Omega$ m. Its mechanical strength varies according to the amount of cold working to which it is subjected.

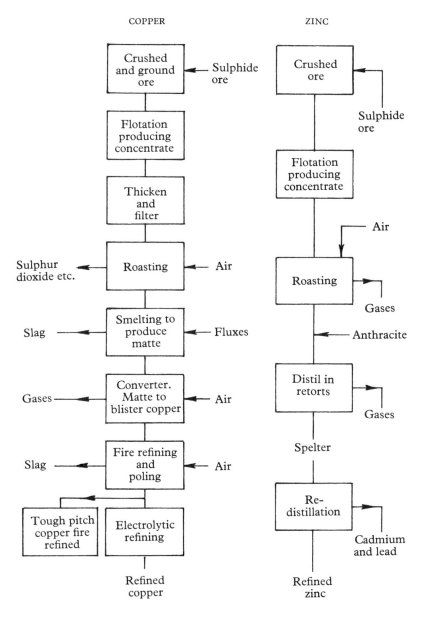

Fig. 5.2   Simplified flow diagrams showing the production of copper and zinc

Hard-drawn wire can have a tensile strength of 500 N/mm² (32·5 tonf/in²). When annealed after cold working the tensile strength is about 232 N/mm² (15 tonf/in²).

The *International Electrotechnical Commission (I.E.C.)* has laid down *Standards of Resistivity*, on which the conductivity of annealed copper is 100 per cent. Oxygen-free copper, which has high thermal and electrical conductivity properties, can exceed this, reaching values on the scale of 106 per cent. This is achieved by adding special de-oxidizers such as lithium to the molten metal in order to remove the remaining minute quantities of harmful elements.

Copper is resistant to atmospheric corrosion because it forms upon its surfaces an oxidized skin which is self-repairing in the event of its being damaged. It is also resistant to many dilute acids, alkalis and salt solutions.

Copper is one of the most expensive of the common metals but its abundance, ease of manufacture, excellent electrical properties, and its widespread use as a base metal in many alloys have made it one of the most valuable metals available to all branches of engineering.

5.4 Forms of supply

Copper from the smelting works is supplied in the form of *blister cakes, wire bars, ingots,* and *billets.* After processing it is supplied as plate, sheet, strip, foil, bars, rod, wire, tube, and various sections

Wire bars are hot rolled into rods of 6-10 mm diameter before being cold drawn through a series of progressively smaller dies until the required diameter of wire is reached. The wire is supplied from 0·025-12·0 mm diameter. Specially shaped wire sections are produced quite easily and the metal can also be of a desired hardness and strength.

Cakes and slabs are hot rolled to produce plate, sheet and strip. Copper in these forms is used in food processing industries, distilleries, chemical plant, domestic hot water tanks, roof sheeting, electrical conductors and earthing terminals, domestic uten-sils, and the numerous small batch parts used extensively in the electrical and tele-communications fields.

Bars and rods are produced by rolling, drawing and extruding. These sections are used for busbars, high-voltage switchgear, electric motor components, bolts, screws, nuts, rivets and nails. Rod is cold drawn copper supplied up to 50 mm diameter.

Plate is flat material over 10 mm thick and 300 mm wide. Various widths can be supplied within the limits of the rolling plant and lengths vary up to 4 m, although pro-vision can be made to produce special lengths.

Sheet refers to material over 0·5 m wide with a thickness range of 0·18-10·0 mm. The usual maximum size of sheet is about 4 m by 1 m.

Strip and foil is produced to the following dimensions: strip is flat material up to 0·5 m wide and above 0·15 mm thick—below this thickness it is foil; strip and foil supplied in rolls can be up to 0·6 m wide; foil below 0·1 mm thickness is very expen-sive due to difficulties of manufacture.

5.5 Copper alloys

Copper is the base metal for a large number of very useful alloys (see Table 5.1). The so-called *silver* coins are alloys of copper and nickel, while the *copper* coins are alloys of copper, tin and zinc.

The various brasses, bronzes and gunmetals are all copper alloys, the brasses being essentially yellow-coloured and the other two alloys golden. Brass is an alloy of copper and zinc; bronze is an alloy of copper and tin plus small percentages of other elements; and gunmetal is an alloy of copper, tin and zinc.

5.5.1 Brass

Brass has mechanical properties which are much better than those of pure copper— it is also cheaper. Many of the ideal fabrication properties of copper are retained and the corrosion resistance is not severely reduced. One main disadvantage of brass is that it is not as good a conductor of electricity as copper, which limits its uses in this respect to small switching components.

Brasses can be tinned without difficulty and can be joined by soft soldering using tin-base solder. The high-copper brasses are better soldering metals than the low-copper brasses owing to the decreased risk of cracking as the copper content increases.

The main problem when welding brass is the vaporization of the zinc. This can be overcome using an oxidizing flame and a flux of the borax-boric type. The brass filler rod used should contain silicon and other special additives which form an oxide layer over the weld and keeps down the tendency to vaporize.

Spot- and seam-welding of the lower zinc-bearing brasses can be carried out successfully.

5.5.2 Phosphor bronze

Phosphor bronze is a copper-tin alloy with a small percentage of phosphorus to act as a de-oxidizing agent, assist the casting characteristics, and increase the properties of hardness and strength.

All phosphor bronzes possess good cold-working properties and have good resistance to corrosion and abrasion. When a small percentage of lead is added the machining properties compare favourably with free-cutting brass.

This alloy is used for making springs and spring clips, electrical contacts, bearing sleeves for lightly loaded rotating and sliding shafts, wire mesh screens and wire 'cloth', steam condenser tubes and components used in corrosive conditions.

5.5.3 Silicon bronze

Silicon bronze is an alloy of copper, zinc and silicon, with small percentages of manganese and iron; or it can be an alloy of copper, silicon and tin. Metals in this range are available in cast or wrought form, mainly for use in chemical processing plant and equipment. These bronzes have very good resistance to corrosive environments and low electrical and thermal conductivities. A small percentage of iron improves the tensile strength and hardness, while tin and zinc increase the fluidity in the mould and help to produce good quality castings.

Silicon-bronze alloys are excellent hot-working metals and can be rolled, forged and extruded with ease.

TABLE 5.1

Typical composition and forms of supply of a small selection of copper alloys

Description of alloy	Composition (%)				Tensile strength (N/mm^2)	Remarks and applications
	Cu	Zn	Sn	Other elements		
Free-cutting brass	58	39	—	3 Pb	386	BS 249 For high-speed machining of repetitive parts. Produced in rod form and a variety of sections
Cartridge brass	70	30	—	—	324	BS 267 A deep drawing brass of high ductility. Used to make cartridge casses, etc. Produced as cold-rolled sheet and strip
Basis brass	63	37	—	—	340	BS 265 A general-purpose brass produced as cold-rolled sheet and strip
60/40 brass	60	40	—	—	370	BS 2870 A good hot-working brass produced as sheet and strip
Naval brass	62	37	1	—	340	BS 409 and BS 1541 Similar to 60/40 brass but with improved resistance to corrosion. Produced as plate, sheet, and strip
Forging brass	60·5	38	—	1·5 Pb	432	BS 2872 This brass has good machining and stamping properties. Produced as plate, sheet, and strip
Phosphor bronze (high tin)	89·6	—	10	0·4 P	1004 when hard	This alloy is used for making springs
Conductivity bronze	98·5	—	1·0	0·5 Cd	278	This bronze is used for telephone wires

5.6 Summary

Copper ores are crushed to a fine powder and floated on a stream of frothy water to separate the copper concentrate from the gangue. Concentrate is either pre-roasted or fed directly to a reverberatory furnace for smelting, the end-product being copper matte. Matte is processed in a converter to oxidize the impurities.

The metal is transferred from the converter to a casting furnace which continually casts blister copper. Blister copper is fire-refined in a reverberatory furnace to reduce the sulphur and oxygen contents prior to making wire bars.

Electrolytic refining produces tough pitch copper having a purity of 99·99 per cent. Copper can be supplied in both cast and wrought forms.

6. ALUMINIUM AND ALUMINIUM ALLOYS

6.1 Raw materials

Aluminium is derived from the ore bauxite, so called because it was first discovered at Les Baux in the south of France. The ore consists of aluminium hydroxide and earthy matter containing other elements. It is first crushed to manageable size, dried, and then treated with hot caustic soda solution under pressure. The aluminium hydroxide is dissolved and sodium aluminate is formed.

The sodium aluminate is diluted with water and allowed to cool prior to being pumped through filter presses to remove the unwanted matter. The filtered material in solution is passed to large tanks and mixed with a small quantity of pure aluminium hydroxide which settles on the bottom of the tanks.

When sufficient of the aluminium hydroxide has precipitated, the contents of the tanks are pumped away to filter out the usable solid material. This is washed free of caustic soda and heated to 1100°C in a cylindrical rotating kiln called a *calciner*. The result of this heating process is *alumina*, a white oxide powder of just over 90 per cent purity.

6.2 Refining

Alumina is changed into aluminium by an electrolytic process. This is carried out in special reduction cells which are carbon-lined mild-steel tanks. The cells are partly filled with electrolyte in the form of molten *cryolite*, a fluoride salt of aluminium, plus other additives which lower the melting point, and this solution is maintained at a temperature just below 1000°C. Alumina is added to the electrolyte to give a 5 per cent alumina solution. Anodes in the form of rods made from carbon are lowered into the electrolyte, the cathodes being the carbon linings of the reduction cells. When an electric current is passed through the anodes to the cathodes, the alumina decomposes and molten aluminium is released which then falls to the bottom of the tanks.

The process is continuous, with alumina being fed to the cells and aluminium drawn off through tap holes and then cast into *pigs*. It requires approximately 2000 kg of alumina, 700 kg of anode material, and 18000 kWh of electricity to produce 1000 kg of aluminium.

A diagram of a typical cell is shown in Fig. 6.1, and Fig. 6.2 shows a simplified flow diagram of aluminium production.

6.3 Properties of aluminium

Aluminium pigs are remelted, the unwanted matter is removed, and the metal is cast into ingots for further processing. The purity of the aluminium at this stage can be as high as 99·8 per cent and this can be improved upon by another electrolytic refining process.

Pure aluminium is a silvery-white metal which is very soft and ductile. Its melting point is 660°C and it has a specific gravity of 2·7, i.e. about one-third that of steel. An

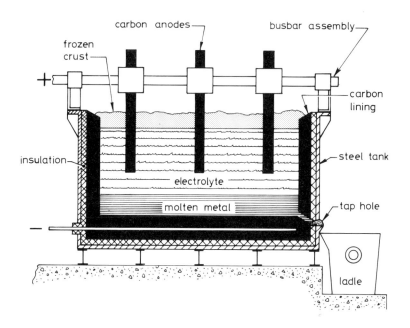

Fig. 6.1 An electrolytic cell producing aluminium

oxide skin forms on the surfaces of aluminium and this gives the metal protection from atmospheric corrosion. The electrical conductivity is second only to copper, it is non-magnetic, and will not spark when struck by a steel hammer. Mechanical strength is low, being about 93 N/mm^2 in the annealed condition, approximately one-third the strength of low-carbon steel.

Pure aluminium is non-toxic and so it is made into cooking utensils and other containers for food and cosmetics. Aluminium foil is used for many applications in the electrical and telecommunications industries.

Aluminium can be cast, forged, stamped, rolled, drawn, extruded and spun. It can be riveted, brazed, welded and glued using organic bonding agents.

Forms of supply of aluminium and aluminium alloys can be as castings or wrought material. The wrought forms are: plate, sheet, strip, and foil produced by rolling; bars and various sections produced by extrusion; tubes and wires produced by drawing; forgings and stampings.

Surface treatment of aluminium products takes two main forms. The metal can be electro-plated or chemical and electro-chemical processes can be used to thicken the naturally-formed oxide coating. The best-known of these processes is *anodizing* in which an electric current is passed through an electric cell in which the aluminium is the anode. The surface produced in this way will take light- and heat-resistant dyes to give a range of attractive colours.

MATERIALS AND PROCESSES

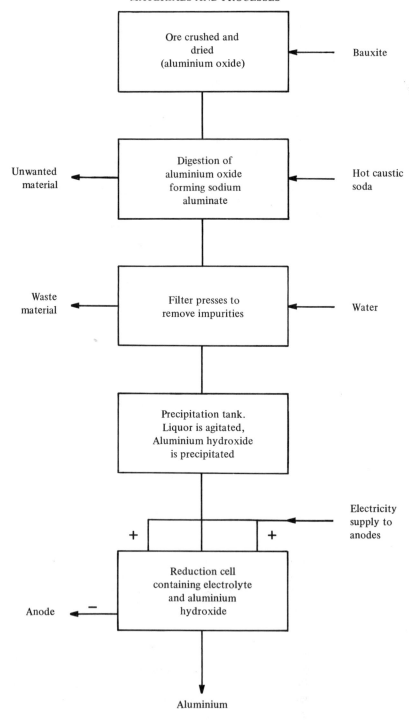

Fig. 6.2   Simplified diagram of production of aluminium

6.4 Alloying elements

Aluminium is alloyed with a variety of other elements, the most important being copper, silicon, zinc, magnesium, iron, manganese and nickel. Many of the alloys will respond to heat treatment and are consequently stronger and harder than the pure metal.

6.5 Summary

The ore from which aluminium is derived is called bauxite. Following a series of chemical treatments, washing, filtering and drying, a white oxide powder called alumina is produced. Alumina is changed into aluminium by an electrolytic process. Pure aluminium has an electrical conductivity value second only to copper. Aluminium can be supplied in cast or wrought forms.

7. MAGNETIC MATERIALS

7.1 Magnetic properties

A *magnetic material* is one that is capable of being magnetized, although the term should be confined to the small group of pure metals that display a magnetic property similar to that of iron. This property is called *ferromagnetism* and exerts a strong influence on magnetic lines of flux, tending to intensify the flux density. The elements, other than iron, tl ⸱t display ferromagnetism are nickel and cobalt.

Pure iron exhibits magnetic properties when placed in a magnetic field, but reverts to its unmagnetized state when removed from the influence of the field. Some steels and other alloys once placed in a magnetic field become *permanent magnets*. Metals which behave in this way are referred to as *soft* and *hard* magnets respectively.

Permanent magnets are used for applications where external power is not available to maintain the magnetic field. The hard magnet will retain its magnetism for an unlimited period provided that it is not submitted to demagnetizing influences. The ability of the material to withstand demagnetization is referred to as its *coercive force*. Soft magnetic materials, which owe their properties to the presence of a magnetic field such as an electric current, demagnetize quickly upon the removal of the external magnetic field and should retain the smallest possible residual magnetism or *remanence*. The highest possible magnetic *permeability* (ease of magnetization) is desirable, coupled with the minimum absorption of energy in an alternating magnetic field.

7.2 Hard magnetic alloys

High-carbon steel in the hardened condition was the earliest of the permanent magnet alloys, and this was followed by high-carbon steels containing tungsten and chromium. Eventually the addition of large amounts of cobalt to high-carbon chromium steels brought to its limit the development of permanent magnet materials based on hardened steel.

The next important discovery was the magnetic properties of alloys containing aluminium, nickel and iron. These were given the name *Alni*, derived from Al and Ni, the chemical symbols for aluminium and nickel. One big advantage was that these alloys were made from less expensive raw materials compared with the chromium-cobalt steels, but in addition they showed considerable improvement in magnetic performance. The coercive forces were higher and the energy content values were improved.

The *Alni* group of alloys formed the basis of a number of important magnetic materials such as *Alnico*, in which copper and cobalt were added to the metal. A method of improving the properties of the alloys still further is to place them in a strong magnetic field during heat treatment. This produces a metallic structure which has *directional* characteristics, that is, a piece of the metal will 'align' itself in the direction of the magnetic field.

Other materials with good magnetic properties are the specially formulated casting alloys which are cast using advanced techniques designed to produce a structure composed of long crystals oriented in the direction of the magnetic field. The disadvantage of these materials is that practical considerations reduce the range of shapes available.

The mechanical properties of hard magnetic materials show that they are also mechanically hard. Some alloys in the Alnico group, for example, attain hardness values that make machining very difficult.

The methods of production are casting and *sintering*, or powder metallurgy as it should be called. The latter method produces a magnet with properties that are almost equal to those of the cast alloys. The techniques involved in powder metallurgy allow for the production of large quantities of small magnets at economic prices. Furthermore, fairly complicated shapes and deep or through-holes can be produced without difficulty.

Permanent magnet alloys in storage will not lose their magnetic properties provided *keeps* are used and there is no close contact with other magnets of similar magnetic force. Complete demagnetization requires the proximity of a field equal in strength to the magnetizing force, which gradually reduces in magnitude.

Permanent magnets are used in loudspeakers of all kinds, television sets, tape recorders, radios, electrical measuring aids, magnetic switches, motors and generators, relays, brakes, railway signalling devices, machine shop holding equipment and magnetic chucks, metallic particle separators, metallic crack detectors, door catches, and cycle dynamo lighting sets.

Table 7.1 shows typical compositions and properties of Alni and Alnico materials.

TABLE 7.1

Typical composition and properties of Alni and Alnico magnetic alloys

Type of material	Approximate composition (%)						Approximate properties		
	Fe	Al	Ni	Co	Cu	Ti	Spec. gravity	Electrical resistivity at 20°C ($\mu\Omega$ m)	Tensile strength (N/mm^2)
Alni	Balance	12	26	–	0·5	0·6	7·0	0·60	100
Alnico	Balance	9	21	13	7	3·5	7·3	0·60	250

7.3 Soft magnetic materials

Magnetically soft materials must possess high permeability and be capable of rapid demagnetization following the removal of the magnetic field, together with the lowest possible retention of residual magnetism. The absorption of energy from an alternating magnetic field must be kept to a minimum.

The earliest soft magnetic material used was near-pure iron which did not possess high initial permeability and magnetizing properties. The addition of about 4 per cent silicon to pure iron not only improved these qualities but also provided high permeability at low magnetizing forces.

The iron-nickel alloys, containing between 50 per cent and 78 per cent nickel, are now widely used because they provide the desirable characteristics of extremely high initial permeability coupled with low field strength, the improvement in initial permeability being of the order of 15 times that of soft iron. These high nickel alloys form the basis of the materials known by the trade names *Permalloy* and *Mumetal.*

The *properties and uses* of high permeability alloys are partly determined by their ductility. They are easily worked into wire and tape as well as thicker sections. These alloys were used to screen underwater telegraph cables and contributed to the production of distortion-free signals and reduced attenuation by making possible the necessary high circuit inductance. This technique was called *magnetic loading*, or more simply, *loading*.

Improvements in telephone efficiency were brought about by using loading coils spaced out at regular intervals along the cables. Initially the coils had powdered iron cores, eventually replaced by the more efficient nickel-iron powders now used in highly congested urban telephone systems.

Nickel-iron sheets are used in transformers for audio-frequency equipment, electrical measuring instruments and power distribution. Magnetic shielding properties are directly related to permeability, so the high permeability alloys Permalloy and Mumetal are used to make magnetic screens for communications equipment and indicating instruments.

Lower alloy materials, such as nickel-iron containing between 35 per cent and 50 per cent nickel, provide materials having maximum permeability and low coercive values. In order to exploit the properties of these materials it is necessary to remove the impurities carbon, oxygen, and sulphur. This is achieved by heat treatment processes in association with hydrogen. Further improvements are made for some purposes, such as higher electrical resistivity, by adding small amounts of copper, silicon and molybdenum.

Alloys in this range containing 45-50 per cent nickel are used as laminations in small power transformers, generators, and in communications equipment where high flux density conditions prevail.

The metals containing about 35 per cent nickel have increased resistivity but reduced initial permeability. These alloys are used for the cores or armatures of relays, solenoids, etc.

Nickel-iron alloys of all types have a good resistance to corrosive atmospheres and are used extensively in humid conditions and underwater equipment.

7.4 Slightly magnetic alloys

Nickel-copper and nickel-iron alloys which sometimes contain manganese or chromium have been developed with special magnetic properties. The physical nature of these alloys is such that they lie close to the dividing line between being magnetic or non-magnetic, that is, the *magnetic transformation point*. The one influence which can cause a change in the magnetic properties is variation in operating temperature.

Materials of this type are used for applications of temperature compensation and control, for example, in domestic and industrial kilowatt-hour meters where the special alloy is attached to one pole of the permanent magnet and acts as a magnetic shunt compensating for resistance variation of the rotating disc as the temperature varies.

Alloys in this range are also used as part of thermal control devices where sharp increases in temperature can cause component damage or fire. The control is effected

TABLE 7.2.

Typical compositions and properties of some magnetic materials

Typical composition (%)		Type and condition of alloy	Properties and applications
Fe	Other elements		
99·9	None	Near pure iron, annealed.	Used for d.c. applications; plungers, solenoids, and pole pieces. High permeability, very low coercive force. Soft magnets.
Balance	0·18 C	Mild steel with usual percentages of impurities, annealed	Motor frames, generator frames, and yokes, d.c. pole pieces. Soft magnets.
Balance	0·50 Si	Silicon iron, annealed	Pole pieces, armatures, relays. Soft magnets.
Balance	3·0 Si	Oriented material produced by annealing and rolling	High permeability and low coercive force. Used for laminations in power transformers of high efficiency. Soft magnets.
Balance	50 Ni	*Permalloy* type Annealed or annealed and cold rolled	High permeability and low coercive force. Used in audio transformers, solenoids in telecommunications equipment, generators, magnetic screens, magnetic amplifiers. Soft magnets.
Balance	5 Mo, 79 Ni	*Permalloy* type, heat treated	High permeability alloy. Used in special transformers and for high-frequency applications. Soft magnets.
Balance	25 Ni, 12 Al, traces of Cu	*Alni - Alnico* type, heat treated	Supplied in cast form and used for permanent magnets. Very hard and difficult to machine.
Balance	14 Ni, 8 Al, 3 Cu, 24 Co	*Alnico*, heat treated	Can be cast or sintered. Very hard and difficult to machine. High-quality permanent magnets.
Balance	20 Ni, 9 Al, 3 Cu, 25 Co	*Alnico*, heat treated	Can be cast or sintered. Very hard. High-quality permanent magnets.
Balance	0·16 C, 0·2 Si, 2 Mn, 8 Ni, 18 Cr	Austenitic steel, heat treated	Non-magnetic steel possessing good corrosion and electrical resistance. Used in circuit breakers, alternators, and switch gear.
Balance	6 Mn, 11 Ni	*Nomag* cast-iron alloy	Non-magnetic or weakly-magnetic.
Balance	20 Ni, 5 Cr	*SG Ni - Resist* alloy cast iron	Non-magnetic.

by an armature which is released from a magnet at a pre-determined temperature and operates an isolating switch or control mechanism.

7.5 Non-magnetic alloys

Some alloy irons and steels are non-magnetic and often have the combined additional properties of resistance to corrosion, erosion, metal-to-metal wear, high temperature environments, and oxidation. They are easy to machine, the casting characteristics are good, they are strong, and in some cases have high strength and ductility over a wide range of working temperatures.

The most notable of the cast irons are the trade materials *Ni-Resist, SG Ni-Resist* and *Nomag*, all of which display a high electrical resistance in addition to being non-magnetic. These materials are used for components in electrical equipment where non-ferrous alloys, which have non-magnetic properties, cannot be used because of their high electrical conductivity. Some of the applications are circuit breaker components, control gear covers, meter and instrument cases, end plates on alternator stators, turbo-generator parts, resistance grids, switch control levers, busbar clamps, transformer covers, insulator adaptor rings, insulator housings, electric furnace electrode holders, and cable-connector boxes.

The 18 per cent chromium, 8 per cent nickel alloy steels are non-magnetic under certain conditions. The steel was developed initially for its resistance to corrosion and moderately high temperatures but the added characteristic of being non-magnetic is sometimes applied with success.

7.6 Summary

A magnetic material is one that displays ferromagnetism, a magnetic property similar to that of pure iron. Pure iron is a soft magnetic material, i.e. it demagnetizes very quickly when removed from an external magnetic field. Hard magnetic materials have the ability to 'store' magnetic force and possess a high coercive force. Soft magnets are usually mechanically soft, while hard magnets are mechanically hard and more difficult to machine.

Soft magnetic materials are used for magnetic screening, transformer sheets, solenoids, and armatures. Hard magnetic materials are used in loudspeakers, television sets, tape recorders, radios, magnetic switches, relays, magnetic chucks, etc. Slightly magnetic materials are used in temperature-sensitive devices. Non-magnetic materials are used for circuit breaker components, meter and instrument cases, resistance grids, insulator housings, etc.

Tables 7.1 and 7.2 give details of a number of magnetic and non-magnetic materials.

8. CONDUCTING MATERIALS

8.1 Conductors

Conductors are materials that provide a path for the passage of an electrical charge and can be either solids, liquids, or gases.

The efficiency of a conductor is measured as a percentage, compared with annealed copper taken to be 100 per cent. This figure is called the *International Annealed Copper Standard* (I.A.C.S.) and is the standard laid down by the International Electrotechnical Commission (I.E.C.) and is based on the low resistivity of the near-pure metal. Of all known substances silver is the most efficient conductor and Table 8.1 shows the relative conductivity of a small number of different materials.

These figures should be used only as a guide to the expected conductivities of the metals. There are grades of copper that give values about 100 per cent; all aluminium does not have 62 per cent efficiency; and the value for mild steel is only an average one and will depend upon the composition of the metal.

Small quantities of impurities in silver, copper, or aluminium will greatly reduce the conducting efficiency. For example, an amount of phosphorus in copper as low as 0·05 per cent can reduce the electrical conductivity by about 25 per cent.

TABLE 8.1
Conductivity ratings of materials

Material	Conductivity (%) (I.A.C.S.)
Silver	108
Copper	100
Gold	72
Aluminium	62
Magnesium	38
Zinc	28
Nickel	25
Cadmium	23
Cobalt	18
Platinum	17
Tin	15
Mild steel	12
Lead	8

8.2 Applications

The primary reason for using any conductor is to allow for the passage of an electrical charge. Various techniques are employed to control that charge by using a material suitable for the task and giving the highest possible efficiency coupled with lowest cost.

Conductors are employed in the generation, transmission, and distribution of electrical power to industry, commerce, and private homes. High-efficiency conductors such as copper and aluminium are used for these purposes. This includes the vast amount of material used in generators, motors, transformers, electro-magnets, solenoids, furnaces, loudspeakers, and the distribution systems in ships, aircraft and motor and rail vehicles.

Conductors used in telephone and telegraph networks must produce signal transmissions with a minimum of distortion or excessive attenuation. In some cases iron, a low-efficiency conductor, can be used to achieve these aims.

Safety devices such as fuses require materials which will break a circuit by melting within a fairly close temperature range when the circuit tends to become overloaded.

The known resistance properties of conductors are used to control the thermal emissions of heating elements in furnaces, ovens and domestic appliances; and where the resistance properties of a conductor are closely related to temperature, the materials are used in temperature indication and control. Conductors with very high melting-points and of high mechanical strength are used as lamp filaments.

Electrodes for welding tools and arc furnaces are conductors which may be required to melt and become part of the weld metal or to sustain an arc at extremely high temperature.

The conductors used in meters and some instruments must be made from a material whose resistance does not alter with changes in working temperature, to ensure that the readings are reliable whatever the surrounding (or *ambient*) temperature may be.

A limited number of conducting materials exist whose expansion coefficients are identical with glass. This enables metal-to-glass seals to be made in electric lamps, radio valves, and switching devices.

Where direct temperature measurement is impossible due to extremely high temperatures, *thermocouples* are used. These are often made from rare-metal alloys and consist essentially of two conductors having different chemical compositions joined together at their ends to form a circuit. When heated at a junction an e.m.f. develops which is used to indicate the temperature.

8.3 Copper conductors

In addition to its property of high electrical conductivity, copper has the additional properties of high thermal conductivity, a good resistance to atmospheric corrosion, ease of working into shape owing to its ductility and malleability, and the ability to be soldered, brazed, or welded.

Conductivity copper comprises oxygen-free high-conductivity copper, tough pitch copper electrolytically refined and tough pitch copper fire-refined. (See Chapter 5). The refining processes are carried out to reduce to the minimum inclusions in the material considered to be impurities. Among these are oxygen, phosphorus, bismuth, arsenic, and lead. The total impurities allowed in the high-conductivity coppers, excluding oxygen and silver, amount to 0·03 per cent, giving these metals a minimum purity of 99·9 per cent.

Tough pitch copper is virtually oxygen-free and is the most widely used conductivity material. The 'pitch' in the name refers to the amount of oxygen remaining in the copper, usually between 0·02 per cent and 0·05 per cent.

Oxygen-free copper is virtually pure copper of 99·95 per cent purity. It is produced by remelting and casting cathode copper of 99·9 per cent purity in an oxygen-free atmosphere of carbon monoxide and nitrogen. For certain applications such as electronic tubes, this grade of copper will make a metal-to-glass seal.

The high-conductivity coppers are used for carrying overhead and underground electricity, telegraph and telephone cables, windings for motors, generators, transformers and measuring instruments, underwater cables, signalling circuits, electrical connectors, railway and trolley bus conductor cables, earthing systems, etc.

Beryllium copper, containing up to 2 per cent beryllium, produces an alloy of increased mechanical strength and hardness after suitable heat treatment. In its annealed condition beryllium copper is soft and can be formed into shape without difficulty. After severe cold forming and heat treatment the tensile strength of certain grades can be as high as 1544 N/mm^2 (100 tonf/in^2).

This class of material is used for the construction of current-carrying springs for use in instruments and resistance welding electrodes. The electrical conductivity varies from about 25 per cent I.A.C.S. when fully hardened, to about 40 per cent I.A.C.S. when the beryllium content is low.

Tellurium copper containing about 0·5 per cent tellurium considerably improves the machining characteristics of copper without adversely affecting the electrical conductivity. The machining properties compare favourably with free-cutting brass, and small repetitive parts for switches, motors, and instruments requiring a good finish and dimensional accuracy are produced from this material. The conductivity rating on the I.A.C.S. scale is about 94 per cent, and the alloy is produced in the form of rod, bar and wire.

Cadmium copper containing between 0·7 per cent and 1 per cent cadmium has an increased softening temperature, improved mechanical strength and toughness and, in the annealed condition, a conductivity rating of about 95 per cent on the I.A.C.S. scale. When cold drawn the tensile strength can be as high as 540 N/mm^2 (35 tonf/in^2). The main application of cadmium copper is for long-span overhead power transmission lines for railways and trolley buses. Table 8.2 shows the conductivities of copper and its alloys.

8.4 Aluminium conductors

Aluminium of the required purity for electrical conductors is specially prepared in order that a maximum resistivity can be guaranteed. The metal has good mechanical strength compared with other metals considered on the basis of its lightness for a given volume. It is a very ductile material and can be supplied in cast form and a variety of sections including wire, rod, tube, bar, sheet, strip and foil. It can be mechanically formed to shape without difficulty and is soldered, jointed and welded (see Chapter 6).

Aluminium has on its surfaces a naturally-formed oxide film which replaces itself immediately after its removal, unless steps are taken to exclude oxygen. The film is a good insulator and must be removed from contact faces when making an electrical joint. However, the oxide film is resistant to most corrosive atmospheres and so exposed conductors such as overhead lines are automatically protected.

Pure aluminium is made into stranded wire for the transmission of electricity from generating stations through overhead cables. It usually employs cores of steel cables to give the stranded aluminium conductors improved mechanical strength.

Steel-cored aluminium (S.C.A.) conductors are also used in various types of

TABLE 8.2

Copper and copper alloys—conductivity grades

Description	Composition (%)		Typical properties		Remarks
	Cu	Other elements	Tensile strength (N/mm^2)	Conductivity (%)	
Cathode copper	99·9	0·001 Bi, 0·005 Pb	—	100	*BS 1035, 1036* and *1861*. High-conductivity copper for castings and high grade alloys. Used to produce oxygen-free and tough pitch high-conductivity coppers
Oxygen-free H.C. copper	99·95	0·001 Bi, 0·005 Pb	216	100 +	*BS 1861.* High-quality conductivity copper used for a wide range of duties in electrical, electronic, and telecommunication work. Properties stated are for material in the annealed condition
Tough pitch	99·9	0·001 Bi, 0·005 Pb	216	100 +	*BS 1036.* High-quality conductivity copper electrolytically refined. Properties stated are for material in the annealed condition
Tough pitch H.C. copper	99·9	0·0025 Bi, 0·005 Pb	216	100 +	*BS 1037.* High-quality conductivity copper fire-refined. Annealed
Tellurium - copper	99·0	0·3 - 0·7 Te	154	85	High-conductivity alloy for casting and in wrought form. Has very good machining properties. In wrought form conductivity can be increased to 98 per cent
Chromium copper	99·0 – 99·5	0·5 - 1·0 Cr	340	82	Heat-treated alloy. Used where high conductivity and high strength are required in both wrought and cast material. Can be used in working temperatures up to 350°C. Heat treated
Silver - copper	99·9	0·03 - 1·0 Ag	216	98	The silver content raises the softening temperature. Used where silver-soldered joints are required, e.g. aircraft work. A casting alloy. Annealed
Copper - nickel - phosphorus	99·0	0·85 Ni, 0·15 P	618	58	An alloy of high strength and moderate conductivity which can be supplied in cast or wrought form. Heat treated
Copper - nickel - silicon	97·0	2·5 Ni, 0·5 Si	478	44	A heat-treated alloy of moderate conductivity but with good wear resistance
Beryllium - copper	98·0	1·8 Be, 0·2 total Co and/or Ni	1235	30	A heat-treated alloy that will give varying properties depending upon the treatment. Conductivity is higher than any other material having similar mechanical properties
Cadmium - copper	99·0	1·0 Cd	278	90	In annealed condition. Strength increases when hardened. Used for overhead conducting wires. *BS 672, 2755*
Conductivity bronze	98·5 – 99·5	0·5 - 1·0 Sn, up to 0·5 Cd	278	70	In annealed condition. Alloy having higher strength than cadmium copper when heat treated but with lower conductivity. Used for overhead conducting wires

stranding for telecommunication lines. The methods of installation and jointing are based on the techniques adopted for power transmission.

The metal is fabricated into television aerials and is widely used as screening and sheathing material for co-axial cables with copper conducting cores. Other applications include aluminium core cable for aircraft use, capacitor elements of aluminium foil, variable condensers for radios, capacitor cans, instrument chassis and frames in die cast form, television and radar aerials and reflectors, power transformer windings, bus-bar clamps, switchgear frames and parts, dry-cell battery cases, etc.

Aluminium alloy conductors are produced from the alloy aluminium-magnesium-silicon, which can be used for purposes similar to the pure metal. Special heat treatments make this alloy suitable for overhead cables which do not require a steel core support. Table 8.3 shows the conductivity grades of aluminium and aluminium alloys.

TABLE 8.3

Aluminium and aluminium alloy conductivity grades

BS No	Al (Min)	Others	Tensile strength (N/mm^2)	Elonga-tion (%)	Max resistivity at 20 °C ($\mu\Omega$ m)	Remarks
		Typical composition (%)			Typical properties	
2627 Wire	99·5	0·05 Cu Si, Fe + Cu not to exceed 0·5	78 - 185	15	0·0280 - 0·0283	For use as overhead conductors. Conductivity rating on I.A.C.S. Scale = 62 per cent. Strength and resistivity depend on condition of metal. First mentioned is annealed.
2898 Bars and sections	99·5	As above	62	25	0·0287	As rolled. For use as busbar material, etc., where high strength is not required
2897 Strip	99·5	As above	100	25	0·0280	Annealed. Higher strength conductivity strip.
			145	3	0·0283	Cold worked.
2898 E91E		0·04 Cu, 0·4 - 0·9 Mg, 0·3 - 0·7 Si, 0·5 Fe	200	10	0·0313	High-strength wrought material of lower conductivity. Supplied specially treated to obtain maximum strength.

8.5 Electrical contacts

When the junction point of two conductors can be separated, either by mechanical or electrical means, in order to disrupt the current flow, the junction point is known as an electrical contact. The positive contact is the *anode* and the negative contact is the *cathode*.

The term includes plug and socket connections which may require frequent disconnection, switch contacts for domestic and industrial use, heavy-duty circuit breakers, small switching devices, rheostats, motor vehicle current distributors, relay contacts, etc.

Contact materials must be chosen to suit the operating conditions and provide the highest possible conductivity coupled with cheapness, mechanical strength, hardness, and the necessary metallurgical properties to resist material loss and breakdown in adverse conditions.

Many contact materials can be made from metals already discussed: copper, brass, bronze, aluminium, etc., while those required to withstand severe arcing and the adverse effects associated with it are usually manufactured from special alloys. Table 8.4 shows the typical properties of some contact materials.

Types of contact are many and varied and a few only are described here. Two-pin and three-pin plug and socket contacts and switching contacts supplied for domestic and industrial uses are made from copper, brass, or bronze. Filament lamp contacts usually consist of solder spots on the base of the lamp pressing firmly against spring-loaded brass plungers in the lamp holder.

Power transmission and bus-bar contacts which are semi-permanent can be clamped or bolted and the materials used are copper or aluminium.

Contact surfaces that are frequently *made* or *broken*, either by direct separation or sliding, are subject to a number of adverse conditions. These include deformation and wear owing to pressure, impact or friction, melting, and the transfer of molten metal from one contact to another owing to high temperature arcing and metallic evaporation caused by the arc.

Tarnishing films of sulphides or oxides form on the surfaces of many contact materials. These films are insulators and must be penetrated to make an efficient contact. Film growth increases with temperature and so the area surrounding points of contact subject to arcing will generally have an increased thickness of tarnish. A white powdery deposit remains on the conductor surfaces when a tarnishing film is decomposed by an arc.

8.6 Circuit breakers

Circuit breakers require the properties of high conductivity and resistance to arcing. In some mechanisms dual contacts are employed, one to cope with the arcing and the other being the main contact. The principal problems are welding and erosion of the contact material owing to the strong arcs generated during switching breaks.

Circuit breakers required to cope with currents of up to 100 A at 500 V are often made of silver-tungsten or silver-molybdenum sintered products to resist any tendency to welding, give mechanical strength, and maintain a good conductivity rating.

When heavy currents at high voltages are being switched, the dual contact system is employed. The main contact member can then be of copper, faced with silver, or a silver-nickel composite material. The arc-resisting member consists of a conducting material, such as copper or silver, made into a sintered composite with either tungsten or molybdenum. Since no alloy is formed from the combination of these elements, each can make its own contribution towards the efficiency of the contact. The copper or silver provides the conductivity while the tungsten or molybdenum provides the strength, hardness, and refractory properties necessary to resist the harmful effects of the arc.

TABLE 8.4
Typical properties of some contact materials

Contact material	Resistivity ($\mu\Omega$ m)	Specific gravity	Melting temperature (°C)	Ductility	Relative cost per unit volume (Silver = 1 approx.)
Silver	0·015	10·5	960	Excellent	1
Gold	0·024	19·3	1063	Excellent	78
Copper	0·017	8·9	1083	Excellent	0·018
Nickel	0·080	8·8	1452	Good	0·051
Molybdenum	0·048	10·2	2625	Fair	0·24
Tungsten	0·055	19·3	3410	Fair	0·66
Platinum	0·098	21·4	1773	Excellent	209
Iridium	0·053	22·4	2440	Good	172
Osmium	0·095	22·6	2700	Poor	—
Palladium	0·108	12·0	1554	Excellent	40
Rhodium	0·043	12·4	1960	Good	184
Ruthenium	0·070	12·2	2500	Fair	62

Contact alloys Composition (%)		Resistivity ($\mu\Omega$ m)	Specific gravity	Remarks
90 Ag,	10 Cu	0·03	10·2	Light-duty contacts
60 Ag,	40 Pd	0·42	11·0	Used for relay contacts at over 100 watts switching power
40 Ag,	60 Pd	0·42	11·3	As above. High-duty contacts
90 Pt,	10 Ir	0·25	21·5	For switching powers from 0 - 1 W
80 Pt,	20 Ir	0·34	21·7	For switching powers from 10 - 100 W
90 Pt,	10 Rh	0·18	20·0	Medium - to high-duty contact alloy
50 W,	50 Ag	0·03	13·2	Sintered contact material, heat resistant
60 Mo,	40 Ag	0·03	10·2	Sintered contact material of high strength

8.7 Carbon brushes

Carbon brushes are contacts used on rotating machinery, and they make the connection between the external supply and the windings of the rotating component of the machine. They provide their own lubrication and remain in contact with moving surfaces such as slip rings or commutators, without rapidly wearing away and with the minimum of friction. In addition, the carbon brush will control any short-circuit current generated in the coils by uncompensated voltage.

The electrical resistivity of brush carbon is higher than the majority of metals and this is one of the main reasons why it is used. The resistance between the brushes and the commutator decreases as the current density increases, causing only small variation in the voltage drop across the contact faces. The polishing action and the surface film developed by the carbon on the commutator are desirable properties, since they assist commutation. Both can be controlled to some degree during the manufacture of the brush material.

Carbon and graphite are made into brushes and anodes used in the electrical, telecommunications and electronic fields. There are different grades of these raw materials but the essential manufacturing processes are the same.

Carbon-graphite brush material, for example, consists of very fine particles of pulverized coke mixed with powdered graphite and pitch. The mixture is compacted and moulded or extruded to the required shape and then baked in a gas-fired furnace. During the baking process the pitch is carbonized and volatile compounds are removed. The graphite in the brush material assists the electrical conductivity and improves the frictional resistance.

Some brushes and anodes are composed almost entirely of graphite, while for certain applications graphite containing metal such as copper can be used.

When graphite-metal materials are made the production methods vary. One way is to mix together powdered graphite and powdered metal to form a sinter. Alternatively the powdered materials can be bonded with pitch and baked. In some cases molten metal is introduced into pores in moulded graphite blocks.

Shunts are flexible copper connectors joining brushes directly to machine structures. They provide a current path from the brushes which is unaffected by any brush movement that occurs. The copper shunt cable is either riveted to the brush or securely embedded in the brush material.

8.8 Relay contacts

Relay contacts in electrical and communications circuits may be required to make and break many times every second and, depending upon the application and the working environment, various materials are available.

The problems that have to be overcome with relay contacts operating rapidly for long periods include sticking of the contact faces, the formation of insulating surface films, and the loss of metal from one contact to another.

Where low arc currents of the order of 0·3 A are experienced, silver-copper alloys can be used. At this current the transfer of metal from one contact to another will be minimal. For non-arcing conditions a tungsten cathode and a platinum anode will operate satisfactorily.

In some telephone and telegraph circuits low-duty relays are required for medium frequencies. The main requirement of the contacts is good conductivity. Bronze can sometimes be used in certain clean atmospheres because it has springing properties in addition to good conductivity. However, the most frequently used material is a high-silver alloy with copper, cadmium, and beryllium. This alloy has the required mechanical hardness and a high electrical conductivity.

Platinum-group metals have the following names: *iridium, osmium, palladium, platinum, rhodium,* and *ruthenium.* Of these, all except osmium can be produced in the shapes required for contact use; the most widely used are platinum and palladium.

This group of metals is expensive, although initial cost can be offset by high performance and long operating life. To reduce production costs some of the metals are suitable for plating on less expensive metals.

For switching loads up to 1 W where light mechanical forces are experienced, gold-platinum or gold-palladium contacts are used. Where resistance from a build-up surface film on platinum or palladium contacts is experienced one of the contacts should be gold-plated.

For switching loads of 1-10W, resistance to metal transfer from one contact to another can be increased by using iridium, rhodium, or ruthenium with a consequent improvement in service life.

Where mercury is the contact medium in a relay it is advisable to use platinum as the solid component because it can be wetted but is not attacked by mercury.

Where switching loads are in excess of 100W and temperatures are not high, silver contacts plated with palladium are often used to counter the problems of tarnishing which occurs at low voltage. In high-temperature conditions platinum contacts would be preferred.

Platinum and palladium are produced as thin strip or wire from which contacts can be made. Techniques have also been developed to provide composite contacts which have thin claddings of these metals or their alloys roll bonded to less expensive materials.

Many applications are to be found of sintered metals used to make contact buttons, especially where the powders consist of metals having extremely high melting points. The advantages of sintering are that the composition of the powders can be accurately controlled and the component can be formed to shape requiring no further machining.

8.9 Vacua, gases, and vapours

The free-flowing passage of an electric current depends upon the conducting substance having within its physical structure a large supply of charged particles (i.e. electrons) that are free to move about. A vacuum does not satisfy this requirement and is an effective insulator. However, it is possible to make good use of an evacuated space by placing in it electrodes which do not require air in order to function. In fact, when metal electrodes or filaments are used it is essential that they operate either in a vacuum or surrounded with inert gas to prevent them oxidizing and burning away.

Alternatively, evacuated envelopes made of glass or quartz can be filled with gases or vapours to provide the necessary charged particles that will ensure the passage of an electric current and emit light.

8.10 Evacuated glass envelopes

All types of electric light bulb, ranging from those used for industrial and domestic applications, to vehicle lighting sets and the humble hand torch, are made by fixing and sealing an evacuated glass envelope into a metal base. There are many variations of this type and size of bulbs but the basic construction, shown in Fig. 8.1, is the same throughout.

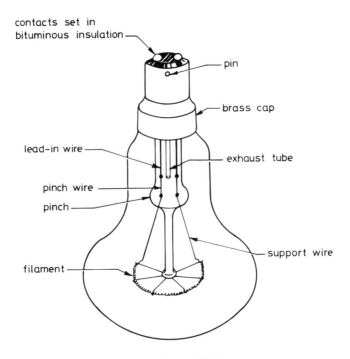

Fig. 8.1   A typical gas-filled lamp

The light-emitting component is made of very thin tungsten wire, much smaller in diameter than a human hair. This *filament*, as it is called, can be either *single-coiled*, or it can be produced first as a single coil, which is coiled again to produce what is known as a *coiled-coil* filament. The reason for this is to enable a long length of wire to be used in a small space and to reduce the losses of heat owing to convection.

The filament is supported by wires made from molybdenum and sealed into a glass stalk called the *pinch* which protrudes into the glass envelope. Attached to the supports are lead-in wires that carry the electric current from the contacts in the end of the bulbs through the support wires to the filament. These wires must expand and contract at the same rate as the glass and maintain a perfect seal at the points where they enter the envelope, so they are made from copper-coated iron-nickel alloys. (These alloys are described in § 10·6, pp 62 and 63).

Tungsten filaments can operate in a vacuum at temperatures up to 2150°C. Above this temperature there is a tendency for the tungsten to evaporate and form a black coating on the inside surface of the glass resulting in eventual failure.

Most electric light bulbs in present-day use are filled with an inert gas but small, low-duty, electric bulbs used in torches and panel lights, etc., have evacuated glass envelopes. Thermionic radio valves, rectifiers, and glow-type switches are similarly constructed.

8.11 Gas-filled envelopes

Tungsten filament lamps with gas-filled envelopes will operate at higher temperatures than those with evacuated envelopes and, since light emission increases with temperature, these are more efficient.

The envelope is evacuated as before but the vacuum is replaced by an inert gas or a mixture of gases such as *nitrogen* and *argon*. The tungsten filament can now operate at temperatures as high as 2750°C, depending upon the size of the envelope, with a consequent increase in luminous intensity.

In special circumstances (e.g. miners' lamps), where high efficiency is required, the gas used is *xenon*. This is very expensive to produce and is not used in ordinary lamps.

The use of metal filaments is limited by operating temperatures, which must increase to obtain greater lighting efficiency. This limitation resulted in the development of *discharge* and *fluorescent* lamps.

Discharge lamps consist of sealed glass or quartz tubes fitted with two electrodes between which an electrical charge is passed through an atmosphere of neon gas, sodium vapour, or mercury vapour. When the electrodes in such a system are not separately heated the tube is called a *cold-cathode* tube. Neon tubes are of this type and have cathodes which are usually cylindrical in shape and made from iron, steel, or copper. A tube containing neon will, in operation, give out light that is an orange-red colour. The name *Neon* has come to mean most tubular lights but gases other than neon are used to give different lighting colours. When helium is used the common white light is seen and when argon is used the light is blue.

Hot-cathode discharge lamps employ separately heated electrodes to increase the electron emission and the envelope is usually filled either with mercury vapour or sodium vapour. So that the discharge can commence at a reasonable voltage, argon or neon are present in the tube. When a current flows the tube absorbs heat, increases the vapour pressure in the tube and enables the mercury or sodium discharge to begin.

Mercury vapour discharge lamps can be of the high-pressure or the low-pressure type. The characteristic colour of these lights is the bluish-green light seen in some street lamps, although in the low-pressure tubes fluorescent powders are used to coat the inside surfaces to give variations in colour. Sodium vapour discharge lamps have a characteristic orange colour and are used to light roadways and industrial buildings.

8.12 Summary

Conductors can be solids, liquids, or gases. Conducting efficiency is measured as a percentage taking annealed copper as 100 per cent efficient. The principal conducting materials in common use are copper, brass and aluminium. Carbon and graphite are used to carry current to rotating machinery because they are self-lubricating and resist wear.

Contacts must be made from materials which will keep arcing to a minimum and resist metal-to-metal transfer. Most glass envelopes for lighting purposes, radio valves, etc., are filled with an inert gas or they are evacuated. Discharge lamps are gas-filled envelopes through which an electric charge is passed.

9. ELECTRICAL RESISTANCE MATERIALS

9.1 Classification

Many electrical appliances, from large furnaces to the humble hair-dryer, require materials for their heating elements which have a high electrical resistance, a resistance to corrosion, and can operate efficiently at elevated temperatures.

In some cases, electrical resistance materials must operate within a temperature range from just above to just below room temperature; for example, where the known resistance of a wire is used in a measuring instrument.

The three principal materials used to manufacture electrical resistance components are: the copper-base alloys, the iron-base alloys, and the alloys based on nickel.

The temperature coefficient of electrical resistance is an important property of a material and is the ratio of the increase of its resistance per degree rise of temperature to its resistance at a defined original temperature.

Electrical resistance materials are *semiconductors* whose specific resistance varies with changes in temperature. With certain materials the resistance decreases rapidly as the temperature rises, and this property can be used to regulate the current flowing in a circuit.

Control devices are made from materials having a negative temperature coefficient, and these are called *n.t.c. resistors*. The method of control is effected by the n.t.c. resistor taking up the excess (or surge) current flowing when the circuit is first energized. Because its resistance is initially high, gradually dropping as the working temperature rises, the n.t.c. resistor protects components such as radio valves whose heaters are cold and have a low initial resistance.

9.2 Copper-base alloys

The alloying elements used in copper-base resistance materials are nickel and manganese with, occasionally, small amounts of iron. This range of alloys is limited in its operations to moderate temperature conditions owing to the lower melting points and resistance to oxidation.

9.2.1 Copper-nickel alloys

Copper-nickel alloys are important because when nickel is alloyed with the copper in increasing amounts the specific resistance rises considerably. At the same time the temperature coefficient of resistance falls, both characteristics having a direct relationship to the increase in the amount of alloying metal.

The alloy called *Constantan*, giving the best combination of high specific resistance and low temperature coefficient, contains 40-45 per cent nickel. The temperature coefficient can be brought almost to zero and, in some cases, to a negative value over a wide temperature range. These materials are ideally suited to the task of protecting electrical, radio, and other devices against possible damage from excessive starting currents. The maximum working temperature of the 40-45 per cent nickel grade of alloys is 400°C, the temperature falling as the amount of nickel is decreased.

A trade material called *Ferry*, used as a control resistance or as thermocouple wire, is in this class and contains from 44-46 per cent nickel.

9.2.2 Copper-manganese alloys

Copper-manganese alloys, which sometimes contain small percentages of nickel, are called *Manganin alloys*. Manganin in certain compositions is used as a standard resistor where its stability under changing conditions of circuit resistance and working temperature make it a high-precision material. It is found in apparatus and instruments that require resistors in practically any form.

For standard resistance use at room temperature, the alloy is varied in order to maintain the desirable temperature coefficient properties. For use at an ambient temperature of 50°C, the manganese content would be reduced to approximately 10 per cent.

9.3 Iron-base alloys

The iron-base electrical resistance alloys are capable of operating successfully at temperatures up to 1400°C.

The principal alloys are composed of iron, chromium, and aluminium, with small percentages of silicon and manganese. Increases in aluminium and chromium in the alloys confer increased resistivity, the aluminium having a greater effect than the chromium, but as a consequence the material also becomes more difficult to work owing to brittleness.

9.4 Nickel-base alloys

Nickel-base alloys are the principal materials used for electrical resistance applications in furnaces, domestic heaters and cookers, and current control devices. The alloys in this range have widely differing properties to suit particular applications but, in general, they have adequate specific resistances and temperature coefficients, good corrosion resistance, and sufficient mechanical strength at temperature to reduce the effects of distortion and sagging. Resistance material can be provided within this range to operate satisfactorily in working temperatures up to 1250°C.

The materials described here will be confined to the trade alloys *Monel* and *Brightray*.

9.4.1 Monel

The Monel range of alloys are used for a number of different purposes, and for electrical resistance use in the form of connecting leads, the alloy is composed mainly of nickel and copper with small percentages of iron and manganese. Its corrosion resistance, moderately high strength, and good ductility make it suitable for leads and connectors to equipment where it may be exposed to temperatures as high as 450°C. The metal can be supplied as rod, tube, wire, strip, and tape and jointing presents no problems.

9.4.2 Brightray

The Brightray series of alloys are composed of nickel, chromium, iron, and silicon in varying percentages and are made to cover a wide range of high-temperature applications, from furnace materials to domestic heating equipment. These materials have the attributes of high strength at elevated temperature, corrosion resistance, and freedom from sagging.

To differentiate between the alloys in the Brightray series and classify them according to their applications, the name for each is followed by a letter of the alphabet.

Brightray C. This grade of alloy is used as the resistance material where switching operations can result in sharp variations of the working temperature. It is therefore suitable for use in electric toasters, radiant fires, cookers, etc. The maximum working temperature is 1150°C and the material gives efficient service over long periods of time. It is supplied in wire or tape form, the tape being rolled from wire. At 20°C it is non-magnetic.

Brightray S. This grade of alloy is made into furnace elements where it will operate for long periods at a maximum temperature of 1150°C. It is supplied in the form of wire and tape and can be welded. At 20°C it is non-magnetic.

Brightray B. A cheaper alloy in the range, Brightray B contains a high percentage of iron and is intended for use as elements operating at temperatures below 950°C. The temperature coefficient of resistance is the highest of the series of Brightray alloys but it is more ductile and is available as sheet, rod, wire, strip and tape. It is used for control resistors and is slightly magnetic.

Brightray F. This grade of alloy is intended for use as heating elements operating at temperatures just below 1000°C. It is ideal for use in furnaces where the atmosphere is either high in sulphur or alternates between oxidizing and carburizing.

At 20°C it can be either non-magnetic or very slightly magnetic, and over its operating temperature range there is a large variation in electrical resistance. However, the temperature coefficient of resistance remains practically constant throughout, making Brightray F a suitable comparator material in temperature measurement.

Brightray H. This grade of alloy can be used as furnace elements for operating temperatures up to 1250°C, retaining high mechanical strength and resistance to corrosion. It is non-magnetic at 20°C and has the highest specific gravity of the Brightray series. It can be supplied as rod, wire, and strip.

9.5 Summary

Electrical resistance materials are semiconductors. Control devices are made from materials having a negative temperature coefficient of electrical resistance. Copper-base alloys are used as resistance materials when the operating temperatures are low to moderate. Nickel-base alloys are used as resistance materials for higher operating temperatures. Resistance materials should possess mechanical strength, high specific resistance, and resistance to corrosion.

Table 9.1 contains a small selection of electrical resistance alloys, indicating typical properties.

TABLE 9.1
Some electrical resistance alloys showing typical compositions and properties

Alloy (All trade names)	Typical composition (%)	Operating temp. (max.) (°C)	Electrical resistivity at 20°C ($\mu\Omega$ m)	Tensile strength (max.) (N/mm²)	Specific gravity (Specific)	Applications and remarks
Ferry	45 Ni, 54 Cu, 1 others	400	0·49	51	8·9	Used for rheostats, instruments, etc., and other low-temp. applications. Low-temp. coeff. of resistance, 2×10^{-5} per °C
Ferry LTC	Impurities reduced to minimum	Properties similar to above				Specially processed to give very low temp. coeff. of resistance below 1.8×10^{-5} per °C between 0°C and 100°C. Supplied as wire
Brightray C	20 Cr, 0·5 Fe (max.), 1·5 Si, 1 others, bal. Ni	1150	1·15	730	8·34	Domestic heating elements and where frequent switching takes place, e.g. cookers, toasters, etc. Non-magnetic at 20°C
Brightray S	20 Cr, 1 Fe (max.), 0·8 Si, 1 others, bal. Ni	1150	1·10	730	8·36	Resistors for furnaces. Easily cold-worked and can be welded. General purpose element to operate under various atmospheric conditions
Brightray B	60 Ni, 16 Cr, 0·3 Si, 1 others, bal. Fe	950	1·10	685	8·28	Low-temperature heating elements for domestic use. High iron content reduces price. Can be used as a control resistance. Non-magnetic at 20°C
Brightray F	38 Ni, 18 Cr, 2 Si, 1·4 others, bal. Fe	1000	1·08	726	7·91	Used in environments where atmosphere is high in sulphur Magnetic properties vary at 20°C
Brightray H	20 Cr, 0·5 Fe (max.), 3·5 Al, 0·8 Si, 1 others, bal. Ni	1250	1·28	762	7·95	Furnace element for use at temperatures up to 1200°C temp. coeff 8.4×10^{-5} per °C. Non-magnetic at 20°C
Monel	68 Ni, 2·5 Fe (max.), 30 Cu	450	0·48	618	8·83	Connecting leads and connectors. Bolts, nuts, and washers. Easily worked and machined, good corrosion resistance

10. BIMETALS AND THERMOCOUPLE MATERIALS

Composite metals or *bimetals* are materials formed by cladding or laminating at least one metal on another without using chemical or electrical deposition.

Bimetal materials are widely used in mechanical, electrical, and telecommunications engineering and the principal reasons for their importance are the lower cost of raw materials and their increased mechanical strength.

Bimetals can be formed in different ways depending upon the metals and the intended applications. The methods of processing available are casting, sintering, pressing, brazing, direct rolling, and what is called the *sandwich process* of cladding.

10.1 Sandwich cladding

This method of producing a metallurgical bond on steel consists of making a *sandwich* of the dissimilar metals, heating it in a furnace and then rolling to the desired shape and size.

Two plates of the required cladding material are placed between two steel slabs, the four edges of which overlap the sandwich metal by a small amount. The facing areas of the cladding are coated with a compound which prevents the sheets from sticking together and also enables them to be parted after treatment. The areas adjacent to the steel slabs are nickel plated so that sound bonding takes place between the dissimilar metals during bonding.

Steel bars are used to fill up the space left by the overlapping slabs and the whole assembly is completely welded along the four edges, to seal in the sandwich material and prevent oxidation during heat treatment. The whole assembly is heated in a furnace to a pre-determined temperature and is allowed to soak at this temperature for a certain length of time.

After removal from the furnace the sandwich is hot rolled to give a metal-to-metal bond of the required thickness and overall dimensions, and then guillotined or flame-cut around the edges to remove the welded-in steel bars. The sandwich is then separated into two steel sheets clad with a desired thickness of a particular metal or alloy, and these can be further heat treated where necessary.

Carbon or alloy steels can be clad in this way and the bonding materials include aluminium alloy steels, copper and its alloys, nickel, and nickel alloys.

Provided that sensible precautions are taken regarding cleanliness, and forming and cutting operations are not excessively severe, clad material can be bent, rolled, machined, sheared, punched, and welded.

10.2 Electrical applications

Copper can be bonded to one or both faces of *steel* in order to provide a cheaper alternative to some of the more specialized copper alloys. High-carbon steel is clad in this way and after heat treatment it can have a springiness similar to that of beryllium-copper, which it can replace for certain electrical contact duties. The electrical conductivity of this bimetal is relatively high when the copper is bonded to both sides of the steel and the material is used for switch blades, fuse clips, contacts, and springs.

Aluminium is bonded to low-carbon *steel* and used to make electron tube elements where it replaces pure nickel at reduced cost but without loss of performance. The aluminium cladding provides a low-density material which reduces the weight of the tube assemblies.

Nickel is bonded to low-carbon steel and is used to make electron tube parts and other electronic components. The nickel is bonded either to both surfaces of the steel or to one surface only. In some cases the steel sheet is faced with nickel on one surface and aluminium on the other. The purpose of the steel-nickel bimetal is to reduce the amount of nickel used and lower the cost of components without sacrificing performance.

10.3 Silver cladding

Silver cladding on copper and copper alloys produces bimetals for the manufacture of electrical and telecommunications contact devices. The silver may be applied to the base metal in a variety of ways. Cladding can be carried out on one or both main surfaces, or on one or both main edges; it can be inlaid to provide a sunken strip running the whole length of the rolled material; or laid on the surface to provide a raised strip.

When required for contact use the cladding material is positioned on an ingot or a slab of the base metal before cold rolling so that when rolled into strip and then blanked out to the desired shape, each piece of base metal will have a contact surface bonded in the required position.

The silver is located where it is needed and there is a reduction in cost by using a bonding process, compared with contacts made from solid silver. In addition the cold rolling work-hardens the materials and increases their mechanical strength without reducing the electrical conductivity.

Silver-clad beryllium-copper is used to make spring contacts for switches and switch blades, while silver bonded to nickel, Monel metal, and steel is made into button-type contacts which are spot welded into position.

10.4 Copper cladding

Copper can be roll-bonded to a number of different metals to provide a wide range of useful bimetals. For electrical and electronic components copper-clad aluminium is of particular significance. The cladding can be carried out as a single-sided or double-sided operation and gives a strong bond. The material can be processed in a work-hardened state or it can be annealed, and is made into sheet, strip, tubing and wire.

Where copper and aluminium conductors meet, this bimetal can be fitted between them in any desired form to provide copper-to-copper and aluminium-to-aluminium joints. This eliminates the erosive galvanic reaction which takes place between the two metals when joined directly together.

The material is used for both electrical and thermal conductivity applications and condenser parts.

10.5 Thermostat materials

All materials respond in some way to changes in temperature, and a small number show sufficiently controlled response to enable them to be used to operate circuit breakers and switching devices at pre-determined temperatures. The operating elements in thermostats are composed of metals having known thermal expansion properties, and these can be either bimetal strips having accurate and controlled flexing characteristics, or rods which have a linear movement.

There are different types of thermostat to suit varying operating conditions and temperature ranges, so the elements themselves must be carefully selected with these factors in mind. Some element materials are bimetal strips while others are alloys in strip or rod form.

The bimetals can be produced by hot rolling composite ingots or slabs and by cold rolling composite strips as described in preceding sections of this chapter.

In most cases the metals selected to make a particular bimetal have wide differences in their thermal expansions in order to produce the greatest possible deflection over the operating temperature range. At the same time it is desirable to have in all operating elements a constant rate of expansion over a range of temperatures.

Classification of thermostat materials is according to their use and the rate and type of movement required of the element. Some elements are designed to have a linear movement and a constant rate of expansion, whilst others are designed to bend or flex at constant or varying rates.

There are *general-purpose* materials which can be used on electrical heating appliances, simple safety cut-outs, and low-temperature domestic equipment such as immersion heaters. *High-temperature* materials are used in domestic cookers, electric irons and small ovens.

Some elements are required to have a *high-activity*, producing a large movement for a small change in temperature. Elements of this type are used in room thermostats to control the programmed setting of a central-heating system. Elements heated by the passage of an electric current and which operate as circuit breakers are called *calibrated electrical-resistance* materials.

Specially developed *corrosion-resistant* element materials have been produced to operate under adverse conditions where other bimetals would rapidly corrode. Although the two component parts of a bimetal taken separately might possess good resistance to corrosion, it often is the case that when bonded together the metals have a tendency to galvanic corrosion in the presence of water or water vapour.

10.6 The Nilo series of alloys

The range of alloys known by the trade name *Nilo* are iron-nickel alloys and provide not only thermostat elements but also metals suitable for making glass-to-metal seals. The name *Nilo* is followed by a number which gives an indication of the nickel content. They have been developed for applications where materials having low and intermediate coefficients of thermal expansion are required. The melting point of the alloys is constant at 1450°C.

Nilo 36. This alloy contains 64 per cent iron and 36 per cent nickel and is also commonly known by the name *Invar*. It is used as the element in thermostats operating at temperatures from 0°C to 100°C. It has a very low expansion rate within this range of temperatures and is used in the control of water heaters and oven thermostats.

Nilo 40 and Nilo 42. Both of these alloys have fairly constant thermal expansions up to 300°C and are used for thermostat elements in domestic cookers and ovens. Owing to their almost straight-line expansion characteristics, these materials allow a linear calibration of the temperature scale to be made. Nilo 42 is also produced in the form of copper-clad wire and is used in electric light bulbs, radio valves, fluorescent tubes, and television tubes, where glass-to-metal seals are required.

Nilo 48 and Nilo 50. Nilo 48 is used as a thermostat material for controlling temperatures up to 450°C. It has a low total expansion over its complete temperature range which makes it suitable for thermostat elements.

The principal uses of both of these alloys is in the formation of glass-to-metal seals in radio valves and electronic equipment, where they meet the essential requirements of a coefficient of expansion matching that of the glass, a melting point higher than that of the glass, good adhesion, and acceptable electrical conductivity.

Nilo 475. This alloy, which consists of 47 per cent nickel, 5 per cent chromium and 48 per cent iron was developed for sealing to the 'soft' glasses used in radio valves, electric light bulbs, television tubes, etc. The thermal expansion characteristics of Nilo 475 match very closely the lead and soda lime glasses used for these applications over a wide temperature range up to 450°C.

Nilo K. This alloy consists of 29 per cent nickel, 17 per cent cobalt and 54 per cent iron. Its thermal expansion matches that of the harder glasses used for high-power valves such as transmitters and rectifiers, electronic components, etc., and it is suitable for use at temperatures up to 500°C.

2. THERMOCOUPLES

A thermocouple is a temperature-sensitive device which consists of two wires of different chemical compositions joined together at their ends to form a circuit. Applying heat to one of the junctions sets up an e.m.f. and a small but measurable current flows. In some combinations of materials forming the circuit, the current flowing is proportional to the temperature difference over a wide range, and this enables accurate temperature measurements to be taken.

A thermo-electric series of metallic elements is shown in Table 10.1, which does not include the list of alloys used to make thermocouple elements. When any two of these metals are selected to form a circuit it is accepted that the current flow set up by the thermal e.m.f. is from the first-named metal to the second. The metals themselves are usually described as A vs (versus) B. For example, when the circuit consists of copper and Constantan, with current flowing from copper to Constantan, this is written as copper vs Constantan.

TABLE 10.1
Thermo-electric series

1.	Bismuth	14	Molybdenum
2.	Nickel	15.	Rhodium
3.	Cobalt	16.	Iridium
4.	Palladium	17.	Gold
5.	Platinum	18.	Silver
6.	Uranium	19.	Zinc
7.	Copper	20.	Tungsten
8.	Manganese	21.	Cadmium
9	Titanium	22.	Iron
10.	Mercury	23.	Arsenic
11.	Lead	24.	Antimony
12.	Tin	25.	Tellurium
13.	Chromium		

Thermocouples are used to measure high temperatures where ordinary thermometers would be unsuitable. These special temperature-measuring devices are galvanometers or potentiometers having in their circuits the special thermocouple materials which not only have the required thermo-electrical properties but are also capable of withstanding high temperatures and oxidizing or reducing atmospheres.

It is often necessary to enclose the hot junction of a thermocouple in a protective metallic sheath, especially where severe atmospheres are encountered. The materials used for this purpose must be very good thermal conductors to ensure accurate temperature readings and to reduce to a minimum the time-lag between operation and temperature indication.

Thermocouple materials include the trade materials *Constantan* and *Eureka* which can be used in conjunction with iron or copper for low-temperature applications. These alloys are copper-base materials containing approximately 40 per cent nickel, and have been supplemented by the similar trade alloy *Ferry* which contains about 45 per cent nickel, (see § 9.2.1).

Thermocouple elements used to measure moderately high temperatures are usually nickel-base alloys called *Chromel* or *Alumel*. Chromel contains 90 per cent nickel and 10 per cent chromium while Alumel contains 94 per cent nickel, 2 per cent aluminium, 2·5 per cent manganese, and small amounts of silicon and iron.

For very high temperature measurement the thermocouple materials are usually made from rare metals of the platinum group and their alloys. The combinations most frequently used are: Pt vs 90Pt-10Rh; Pt vs 87Pt-13Rh; 95Pt-5Rh vs 80Pt-20Rh; 90Pt-10Rh vs 60Au-40Pd; Ir vs 60Ir-40Rh and Ir vs W. The numbers are percentages of the particular element in the alloy.

Figure 10.1 shows graphs of some thermocouple materials indicating the operating temperatures plotted against approximately thermal e.m.f.s.

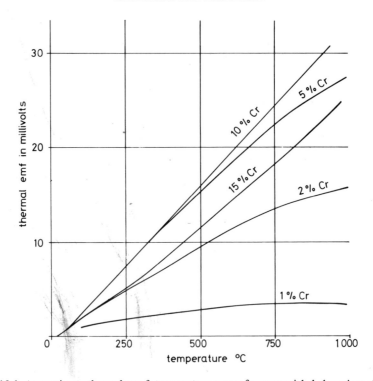

Fig. 10.1 Approximate thermal e.m.f.-temperature curves for some nickel-chromium alloys

10.7 Summary

Bimetals are produced mainly by roll bonding one metal with another or sintering powdered metals. Cost savings can be made by cladding inexpensive metals with more expensive metals. Materials having the desired electrical properties can be correctly positioned on the base metal prior to rolling. Bimetals are used to operate switching devices such as thermostats by making use of the materials' controlled response to changes in temperature. Where operating temperatures are too high for ordinary thermometers, thermocouples are used.

11. ELECTRICAL INSULATING MATERIALS

Materials used as electrical insulators have a high resistance to the passage of an electric charge provided that certain operating conditions, especially that of temperature, are not exceeded. There is a very wide range of insulators having different mechanical and physical properties and so care must be exercised in choosing the correct material for any particular application. For example, impregnated paper is a good insulator and is used for certain power and telephone cables where the operating temperatures are very low, whereas high-temperature electrical insulators are made from glass or other refractory materials.

11.1 Paper insulation

Wood fibres and manila fibres are the principal raw materials used in the production of electrical insulating papers. Wood papers are the most widely used owing to superior quality and lower costs. Paper insulation can take the form of pulp, tape, or string and as a space-filling material in some cables where its electrical insulation properties are not of prime importance.

The disadvantages of paper are its low mechanical strength, low temperature properties, and that it is *hydroscopic*, i.e. it will absorb moisture from the air when in its unimpregnated form.

Power cable insulators are produced by passing each conductor through a paper-lapping machine in which paper tape is spiral wound on to it, attaining a thickness of up to 20 mm depending upon the type of cable. Some impregnated papers operate at temperatures above $60°C$ and stresses approaching 10^7 kV/m.

Telephone cable insulation composed of paper is made by wrapping each wire of a cable separately before enclosing the wires in a sealed, waterproof sheath of lead which protects the paper from moisture and physical damage. Initially, each wire is helically would with paper string, leaving a space between the windings. This is then covered with a layer of paper tape which forms a tube not in contact with the wire owing to the presence of the paper string.

For small diameter conductors a single layer of paper is helically lapped directly to the wire and the whole assembly is then passed through a die of the required diameter. The paper is crushed and creased during this process; the creases provide air space between the conductor and the insulator and also help to hold the paper in position on the wire.

Impregnated paper is used in a wide range of insulators, the impregnating medium acting as a moisture-repellent and a means of preventing natural deterioration over a long period of time. Impregnants are often viscous (thick) oils with or without additives, low-viscosity oils and petroleum waxes. The paper can be supplied impregnated and ready for lapping, or impregnation can take place after lapping, when the completed cable is immersed in oil or dissolved wax.

Sheathed cables containing paper-insulated conductors are sometimes impregnated after they are laid, the insulating oil being introduced at low pressure into the air spaces between the cable sheathing and the paper insulation.

11.2 Cambric insulation

Cambric is a cloth woven from cotton fibres which passes through a number of processes before being varnished and wound into rolls. There are two main varieties of this material in common use, black varnished cambric (B.V.C.) and yellow varnished cambric (Y.V.C.). Both are manufactured in the same way and, apart from the obvious colour difference, the black varnished cambric has better water-repellent properties, a smoother finish, and a higher electric strength than the yellow varnished cambric.

Black cambric insulation owes its colour to bitumen which is added to the varnish. It is used for cable insulation in preference to Y.V.C. owing to its improved electrical properties. Its main advantage over paper insulation is that it does not require fluid impregnation to prevent moisture absorption.

11.3 Insulating oil and wax

Insulating oils produced for the electrical and telecommunications industries are derived from crude mineral oils which must be low in sulphur, waxes and asphaltene. The two principal types are *paraffinic oils* and *naphthenic oils*.

11.3.1 Insulating oils

Insulating oils are distilled from suitable blends of crude oils during the normal processes at an oil refinery. Initially, distillation takes place at atmospheric pressure to remove the so-called volatile fractions, petroleum, kerosene, etc. The remaining oil is heated in a vacuum chamber and the vapours given off pass to a tall cylindrical vessel called a *fractionating tower*. As the mixed vapours rise they cool, the various substances in the vapours causing condensation to occur at different levels in the tower. Each level produces a *fraction*, the heavier fractions at the base and the lighter fractions at the top.

The insulating oils are produced close to the base of the tower, but still contain unwanted matter which must be removed. To achieve this, the oil is mixed with sulphuric acid which attacks the impurities and forms a sludge that is later separated from the oil in a centrifuge. After washing with a mixture of substances to remove much of the remaining acid and sludge, the oil is filtered through various clays where it leaves behind all the remaining unwanted matter, including any asphaltene.

Transformer insulating oil, used in both transformers and switchgear, also acts as a coolant, dissipating the heat generated within the equipment.

Transformer oil should, ideally, have the following characteristics:
(1)   a low viscosity, to enable it to circulate freely between the windings and through the passageways of the transformer;
(2)   a low volatility, to prevent too great a loss of oil through vapourization under normal working conditions;
(3)   the ability to conduct away the heat generated within the transformer;
(4)   a low pour point, so that in very cold weather the oil will still flow freely and not become semi-solid;

(5)   a high degree of chemical stability to resist chemical changes such as oxidation which causes the formation in the oil of harmful sludge, acids, and water;

(6)   a flash point sufficiently high as to constitute no real danger;

(7)   freedom from contaminating elements such as dirt, dust, etc. Care must be taken to ensure that storage facilities are clean and the transformer itself should be so designed that the minimum amount of atmospheric dust is introduced through the breathing system.

The relevant British Standard for insulating oil used in transformers and switchgear is *BS 148:1959*.

Switchgear insulating oil is used to carry out three principal tasks:

(1)   to quench the arc that flashes across certain contacts when a circuit is broken;

(2)   to provide a cooling medium during normal working;

(3)   to insulate the working parts of the circuit breaker.

The most important requirement is to quench the arc as quickly as possible so that complete combustion of the oil cannot result. Because arcing takes place beneath the surface of the oil, a small quantity of the liquid local to the arc is destroyed owing to the high temperatures generated. One result of this is the formation of tiny carbon particles which cause a gradual deterioration of the oil related to the number of times the circuit is broken. In addition, the carbon particles settle and gradually build up along the horizontal surfaces of internal insulators, forming a track which could cause a short circuit.

Circuit breakers operate at temperatures below those associated with transformers and so the insulating oil will be subjected to only moderate temperature-rises—about 40°C under normal operating conditions—compared with the possible 60°C in some transformers. The oil used in most switchgear is similar to transformer oil and conforms to *BS 148:1959.*

Bus-bar chambers can be filled with a high viscosity oil although the more usual filling is a compound based on bitumen. The main purpose of the filling medium in such chambers is the exclusion of dirt, dust, and moisture.

11.3.2 Petroleum wax

Petroleum wax is produced as a precipitate resulting from the solvent treatment of residues which collect during the distillation of petroleum. The precipitate contains unwanted matter which is removed by further solvent treatment followed by filtration through clays. A certain amount of petroleum wax is obtained by natural precipitation during the storage periods of some crude oils. This is refined in the same way as the wax obtained from the distillation of petroleum.

Insulating wax is used as an impregnating material for paper and cloth insulation, as a dipping medium, and as a directly-applied coating on conductors.

11.4 Elastomers

The term *elastomer* describes the range of materials both natural and synthetic which display characteristics normally associated with natural rubber, for example, the ability to extend longitudinally for considerable distances without fracture, and having a high degree of resilience.

11.4.1 Natural rubber

Natural rubber is produced from the sap (latex) of a tree called *Hevea Brasiliensis.* Latex is a milky-white liquid composed partly of water and is coagulated by adding to it formic or acetic acid. The acid acts upon the latex in much the same way as it would upon milk, forming a thick coagulum which floats on the watery element. The coagulated latex is skimmed off and processed through a series of heavy rolls which compacts the latex and squeezes out any remaining liquid. The material that leaves the rolls is washed and then dried in an atmosphere of hot air and wood-fire smoke to produce *ribbed smoke sheet.*

Natural rubber sheet in crude form is a low-strength, low-temperature material of little practical value. It is therefore ground into small particles and compounded with other substances, including inert filler materials such as carbon black or french chalk and used as a moulding powder.

Sulphur is added to rubber compounds to change the structure of the material and enable it to be vulcanized. When about 5 per cent of sulphur is used in a compound which is then masticated, rolled, and subjected to a heating process, the vulcanized rubber that emerges will be soft-textured. As the sulphur content increases the rubber becomes harder until, at over 30 per cent, a hard solid substance called *ebonite* is produced. Ebonite is usually considered to be a plastics material and although it is extruded into rod form, and other sections, it can also be ground to a powder and moulded to make vehicle battery cases and electrical insulators.

Components produced from rubber are moulded to the required shape; the moulding temperatures melt the rubber so that it fills the moulding cavities and then curing or vulcanizing takes place in the mould under heat and pressure.

Cable coverings and rubber insulation applied directly to individual wires are extruded, the material to be covered passing through the centre of a die as the rubber is extruded around it. Some components which require to be completely sealed are dipped into whipped latex compounds and then passed through an oven where vulcanization takes place.

Natural processed rubber is attacked by oils, petroleum and spirit solvents. Its chemical structure also gradually deteriorates, or *ages* when exposed to heat and sunlight, although chemical agents are used to delay this breakdown process.

When cables consist of vulcanized rubber insulated conductors enclosed in a tough vulcanized rubber outer casing they are called *tough rubber sheathed* or *T.R.S.* cables. Similarly, the use of *vulcanized rubber insulation* has given rise to the widespread adoption of the initial letters *V.R.I.* to denote this type of insulation.

11.4.2 Butyl rubber

Butyl rubber is a synthetic rubber produced from the gases isobutylene and isoprene. Unlike natural rubber, it resists the harmful effects of sunlight, rapid temperature changes, and many organic and inorganic solvents. Its chemical structure does not break down with age, and these attributes, coupled with an excellent electrical resistance, make butyl rubber an ideal replacement material for natural rubber in the covering of power cables and for other uses where conditions would be harmful to natural rubber.

11.4.3 Butadiene-styrene rubber

Butadiene-styrene rubber was produced in vast quantities by the plastics industry of the United States of America during World War II as a rubber substitute. It was called *GR-S*, initials which stand for *Government Rubber-Styrene*.

Production is based on the compounding of the raw materials butadiene and styrene, together with chemicals and carbon black filler similar to those used in the production of natural rubber. Small quantities of natural rubber are also used in the compounding in order to produce characteristics that are close to those of the natural product. Sulphur is added to enable vulcanizing to take place and to vary the hardness of the finished product, which has a good resistance to abrasion but lower mechanical strength and elasticity than natural rubber.

Butadiene-styrene rubber is used as a covering for conductors and also as an outer sheath, usually in conditions where it will not come into contact with oils and solvents to which it has little resistance.

11.4.4 Neoprene

Neoprene is made from the chemical chloroprene and is the name given to a number of synthetic materials based on this chemical that are rubber-like but have some useful properties not possessed by natural rubber. Other names by which this material is known are *Polychloroprene* or *P.C.P.* and *GR-M*, the initials used during World War II which stand for *Government Rubber-Monovinyl* acetylene.

Neoprene used in the manufacture of cables is similar to GR-M, is highly resistant to oils and some solvents, and possesses a reasonable mechanical strength and good resilience. It is also resistant to the effects of ozone, sunlight, and heat and will not support combustion.

It can be vulcanized using chemicals other than sulphur, but it cannot be processed into a hard vulcanite in the same way as natural rubber. The electrical insulation properties of Neoprene are not as good as those of natural rubber and so it is used as a sheathing material for rubber-covered conductors. In this way the electrical resistance properties of rubber are combined with the fire-retarding properties of Neoprene. Cables of this type are used in ships. Aircraft wiring consists of glass-braided conductors sheathed with Neoprene.

The principal filler material used in Neoprene is carbon black which, in cable sheathing compounds, may account for almost 50 per cent of total volume of material. In addition to giving Neoprene its black colour, carbon also helps to give it the properties of high resistance to abrasion and cutting.

11.4.5 Silicone rubber

Silicone rubber is unique in being the only synthetic elastomer with an inorganic chemical structure based on silicon and oxygen. Some grades also use carbon so these can be considered as partially organic. Despite the extraordinary behaviour of silicone rubber under conditions which would destroy natural rubber, it is very difficult to distinguish one from the other simply by appearance and feel when they are both the same colour.

Various inert filler materials can be used in the compounds and these contribute to the range of colours available. Certain chemical ingredients are also included to allow vulcanization to take place. In the fully cured condition silicone rubber is chemically inert and can be used in environments where other materials would create a certain risk of contamination.

A trade material called *Silastomer* is widely used in Great Britain and exhibits the following characteristics: it does not melt or char when exposed for short periods to temperatures of 300°C; it can be used continuously in temperatures up to 180°C; and it decomposes at about 400°C. At the other end of the temperature scale the same material shows little or no stiffening and no loss of essential properties to −60 °C.

Although *Silastomer* swells when in contact with some solvents, the material will recover its original dimensions when the solvents are removed, and during solvent contact the properties of the material are not reduced.

Silicone rubber is more expensive than the other elastomers, consequently it is chosen for particular applications where its unique properties can be fully utilized. It is used as a cable sheathing in aircraft electrical wiring and in conditions where high and low temperatures would rule out other elastomers.

Silicone compounds other than elastomers are produced in the form of varnish for insulation use with glasscloth, mica, and asbestos. Silicone materials are used to bond mica into sheet and tape form and also in the production of mica-glass tapes used as high-grade insulators.

11.5 Plastics materials

There are two principal types of plastics material, *thermosetting* and *thermoplastic*. Thermosetting materials cannot be remelted once formed into shape, while moulded thermoplastic materials will melt upon the application of heat. Some plastics materials, such as certain grades of epoxy resins, can be either thermosetting or thermoplastic depending upon the chemical constituents.

All plastics are synthetic products, the majority of the chemicals being produced as by-products of the petroleum industry. The elastomers described in preceding sections of this chapter are plastics materials but have been considered with rubber because they possess similar characteristics.

11.5.1 Vinyl plastics

Vinyl plastics are important insulating materials. One of the most widely used is *Polyvinyl chloride* or *P.V.C.* P.V.C. is a class of materials rather than a single product. It can have the appearance and feel of rubber and exhibits the elastic properties associated with elastomers, but it is not vulcanized. P.V.C. is a synthetic plastics material produced as a white powder from the raw materials acetylene and hydrochloric acid. Chemicals called *plasticizers* are used during the compounding of the powders to give the P.V.C. its degree of plasticity.

The basic material can be produced as powders for injection moulding and extrusion or blended and hot rolled into sheet form. It is a thermoplastic material, that is, it is formed into shape under the action of heat and pressure and can be softened or remelted by applying heat.

Plasticized P.V.C. is colourless and almost transparent and is capable of taking a wide variety of colouring pigments which hold fast over a long period of time. This makes the colour coding of wiring systems easy because the insulating sleeve of P.V.C. will be coloured for its complete length. Plasticized P.V.C. is used as a sheathing material for conductors covered with P.V.C. itself and many other forms of insulation. It is soft and extremely flexible and possesses good dimensional stability. It has a resistance to water and moisture absorption, many acids and alkalis and most common solvents. Although there is a little deterioration of colour when exposed to light and ageing manifests itself by a loss of flexibility, these disadvantages are not experienced until a considerable period of time has elapsed.

P.V.C. is difficult to ignite and burns with some reluctance, with pitting and charring in the area directly exposed to a flame. However, the insulating properties fall away rapidly as the temperature rises, making it unsuitable as an insulator where the temperature is likely to exceed 70°C. Although its electrical insulation properties are good, it is not as efficient as tough rubber where high frequencies are experienced. The current rating of P.V.C. is the same as that of T.R.I. for circuits carrying up to 660 V owing to the fact that rubber has an insulation resistance greater than is actually required for such circuits.

P.V.C. insulation is not as efficient as polythene (§ 11.5.3) but it has a better resistance to abrasion and is tougher. In the telecommunications field its principal use is in the internal wiring of telephone switchboards, jumper wires and exchange bus-bars. For some applications the P.V.C. is covered with cotton braiding which is then treated with a flame-retarding lacquer.

11.5.2 Polyvinyl chloride-acetate

Polyvinyl chloride-acetate (P.V.C.-Ac.) is a thermoplastic material consisting partly of vinyl acetate and partly of vinyl chloride. P.V.C. is a tough plastics material which melts at about 170°C and is resistant to many solvents and chemicals, whilst P.V.Ac. softens in hot water and dissolves readily in common solvents. The new material formed by chemically linking the two together will possess some of the properties of each one. A common formulation contains 85 per cent P.V.C. and 15 per cent P.V.Ac.

P.V.C.-Ac. is a good electrical insulator particularly in the low frequency range and is usually produced for this purpose in the form of injection mouldings. P.V.C.-Ac. with a high acetate and low chloride content makes an excellent varnish medium.

11.5.3 Polyethylene

Polyethylene is a wax-like thermoplastic material more commonly known by the trade name *Polythene*. It is produced from the gas ethylene, a by-product of the petroleum industry, which is heated to about 150°C under tremendous pressure. It has a specific gravity of 0·93, making it lighter than an equal volume of water; it melts at about 110°C; and possesses a good resistance to impact and an excellent resistance to an electric current. It is unaffected by water, acids, alkalis, and many organic compounds, and is tasteless and odourless. Polyethylene is a clear, colourless material that can take a wide range of colouring pigments.

In cable manufacture Polythene is used for high-frequency radio and telecommunications purposes, including distribution cables. It can be used in the form of sheathing, and P.V.C. sheathed polythene-insulated conductors are the first choice in loudspeaker circuits and other audio-frequency applications.

Wire-armoured, low-voltage underground mains cables are making increasing use of polythene as an insulator. Apart from its excellent electrical properties which make it an almost perfect dielectric, its light weight and flexibility contribute to its popularity.

Submarine cables are also making use of polythene which is taking the place of the more traditional insulating material *gutta-percha.*

The disadvantages of using polythene are that it is gnawed by rodents when buried directly in the ground; it is attacked by ultra-violet light unless protected by a sheath of black material such as butyl rubber and, because it will sustain combustion, it cannot be used where there is a possibility of arcing.

Polythene can be moulded, extruded and machined without difficulty and, although the surface of the material is soft and easily marked, it is a tough material which maintains a high degree of flexibility over a wide range of temperatures. Its specific resistance at $20°C$ is 10^{15} Ω m.

11.5.4 Polytetrafluoroethylene

Polytetrafluoroethylene is more commonly known by the initial letters P.T.F.E. It is an expensive plastics material which starts life as a gas prepared from the raw materials fluorspar and chloroform. The granules of P.T.F.E. which result from chemical reactions under high pressures and temperatures are white and have a wax-like feel and appearance.

P.T.F.E. cannot be moulded in the usual way but is sintered, a heating and pressing technique which fuses together a mass of tiny particles. It can be extruded as an insulating sleeve over conductors for high-frequency electrical work, radio-frequency applications, high-temperature aircraft wiring such as gas turbine ignition cables, thermocouple leads, high-temperature transformers, and motors.

Its properties and potential uses are outstanding. The working temperature of P.T.F.E. extends from $-70°C$ to above $250°C$, it is extremely tough yet flexible, self-lubricating with excellent non-stick properties, unaffected by water, will not burn, has no known solvent, does not break down with age, is unaffected by light, and is an excellent electrical insulator of high mechanical strength. Its specific resistance at $20°C$ is 10^{15} Ω m (the same as polythene) and its specific gravity is $2·2$.

Copper conductors sheathed in P.T.F.E. should be protected by a layer of nickel or silver because the extrusion processing temperatures cause severe oxidizing of the metal.

Other fluorine-containing plastics have been produced which combine the remarkable properties of P.T.F.E. with greater ease of working. One such material, called polytrifluorochloroethylene or P.C.T.F.E., can be compression moulded or injection moulded.

11.5.5 Polystyrene

Polystyrene is a colourless, clear thermoplastic that will accept a wide range of colouring pigments. It is produced from the controlled chemical reactions of the raw materials ethylene and benzene.

Polystyrene is one of the best materials available for injection moulding, producing hard, brittle components which have a specific gravity of $1·05$ when no pigments are used. It has excellent electrical resistance properties but, because it is not a flexible material, it cannot be used as a cable covering except when it is compounded with other formulations (see § 11.5.3).

A high-impact polystyrene is produced which is moulded into electric drill casings, small insulators and cable carriers, refrigerator parts, radio cases, domestic electrical appliances, electronic calculator cases, etc.

Polystyrene mouldings have good dimensional stability, are resistant to moisture and a number of dilute acids and alkalis, but attacked by petroleum products. Solvent adhesives are used to join polystyrene to itself and other materials. This plastics material is odourless and tasteless and will soften in boiling water, so it is used only in low-temperature environments. When dropped on a hard surface it emits a metallic ring.

In the form of sheet, film, rod, and tube polystyrene is employed as an insulator in high-frequency equipment such as radio and television where its non-tracking properties can be used to good effect.

11.5.6 Acetal resin

Acetal resin is derived from formaldehyde and ethylene. It is a thermoplastic manufactured in Great Britain by I.C.I. under the trade name *Kematal.*

Components moulded from acetal resin have a long and useful life in air at temperatures up to 100°C. This plastics material offers good resistance to many chemicals and is also resistant to strong alkalis, petroleum, detergents, and many organic salts but it should not come into contact with strong acids, phenol, or acetone and other solvents.

It is a good moulding material provided that care is taken to control the processing temperature. The mouldings possess very good electrical resistance properties and can be used with high voltages. Transformer coil formers, terminal blocks, and switchgear components are moulded from acetal resins.

11.5.7 Phenol formaldehyde resins

Phenol formaldehyde (P.F.) resins are produced from the raw materials phenol (carbolic acid) and formaldehyde reacted together with other chemicals under controlled conditions. The result is a powder to which filler materials are added before the product is ready for use. The mouldings are thermosetting and are produced in compression moulds or transfer compression moulds to form components such as switch box covers, three-pin plug bodies, adaptors, motor vehicle electrical distributor covers, radio cabinets, electric light fittings, electric smoothing iron handles, etc. Metal parts such as brass and copper conductors and earthing points can be moulded in place.

Filler materials play an important part in determining the characteristics of the moulded components which, in general, are hard and brittle. A common commercial grade of powder would contain a high proportion of woodflour, a very fine dust produced from hard woods. Mica dust filling is used to improve the electrical resistance properties but one of the best all-round powders contains chopped paper. Its electrical resistance is good and this is combined with a reduction in brittleness. P.F. resins are limited as to colour, those commonly seen are dark brown and black. The specific gravity of filled P.F. resin is about 1·4; it is resistant to attack by alcohol, oil, most common solvents, weak acids, and alkalis but it does absorb water which can reduce its efficiency as an insulator.

11.5.8 Phenolic laminates

Phenolic laminates are used in radio and television receivers, coil formers, panels, base boards, printed circuit boards, and transformer bushings. At a particular stage in its production, phenol formaldehyde resin is a syrupy liquid and can be thinned using quick-drying solvents. The thinned resin is used to impregnate woven cotton cloth, asbestos cloth, glass cloth, or paper.

Following a drying process these materials are cut into suitable lengths to form sheets which are stacked between the heated platens of a press and pressure is applied. The combination of heat and pressure 'cures' the resin, the sheets form a single panel, and upon removal from the press a strong, tough, hard and impervious material is produced which is commonly known by the trade names *Tufnol* or *Paxolin*.

There are many types of Tufnol and Paxolin resin-impregnated laminates, all of which are electrical insulators, but the best materials for this use are those based on paper.

The laminate is mechanically strong and yet can be machined or punched to shape after being heated slightly. It can be supplied as sheet or tube and is capable of withstanding continuous working temperatures of 80°C. The thickness of laminated sheets varies from 0·4 mm to 76 mm and the over-all size of sheet is approximately 1 m square.

11.5.9 Urea formaldehyde

Urea formaldehyde (U.F.) thermosetting plastics materials are similar in many respects to phenolics. Urea is produced from the gases ammonia and carbon dioxide and reacted with formaldehyde under controlled conditions to form a colourless syrupy resin. Filler materials similar to those used with P.F. resins and colouring pigments are added to the resin, producing a soft crumbly mass which is dried and prepared for moulding.

U.F. mouldings can be coloured; they are heat-resisting, odourless, and tasteless and are unaffected by oils, alcohol, petroleum, and most common solvents. The operational temperature range is from − 50 °C to 80 °C, and they burn only with difficulty. The electrical insulation properties are good, with a resistance to arcing and, although U.F. materials will absorb moisture, the electrical resistance properties are little affected.

Resins are used as impregnants for a wide range of different materials and also form the basis of extremely efficient adhesives. Laminated materials are produced similar to those using P.F. resins but, because pigmenting is possible, the U.F. laminates are more decorative.

The moulding powders are processed to make industrial and domestic switches, plugs and sockets, light fittings, and small insulators, all of which can be produced in a wide variety of attractive colours. The coloured electrical fittings in the modern home or factory are almost certain to be made from urea formaldehyde moulding powders.

11.5.10 Polyester resins

Polyester resins are formed by compounding a number of acids, glycols, and styrene together under rigidly controlled conditions. When mixed with special chemical hardeners the compound forms a very hard thermosetting material.

Large quantities of polyester resin are used in the production of glass reinforced plastics (G.R.P.), commonly but incorrectly called *glass fibre* or *fibreglass*.

When used to impregnate woven glass cloth, polyester resins form a strong and rigid construction material having a wide variety of uses. The electrical resistance of G.R.P. is good, with a typical resistivity value of $10^{11}\,\Omega$m and the material is suitable for use in the open air because it does not rot, absorb moisture, or break down due to age. Thin G.R.P. boards are used as a base for printed circuits.

Certain polyester resins are most suitable for *encapsulating* or *potting* delicate electrical and electronic components. The component is dipped into the liquid resin which is allowed to cure. This gives a hard impervious skin protecting the component from moisture, dust, oxidation and sometimes vibration. Heat-sensitive electrical devices can be protected in this way. The electrical connexions to any potted component are completely sealed in and the contact surfaces are kept free from resin by using a special release agent which is cleaned off after the resin has cured.

11.5.11 Epoxide resins

Epoxide resins are produced in Great Britain by CIBA (ARL) Ltd., under the trade name *Araldite*. Sometimes known as *epoxy* resins, this class of plastics materials is made from the basic raw materials epichlorhydrin and dihydric phenols. Other chemicals are used in the formulations, together with any suitable fillers.

Immediately prior to use the resins are mixed with active chemical curing agents and the result is a sticky, milky-white paste which after a time changes to an amber-coloured, hard and smooth thermosetting solid.

Epoxide resins are used to encapsulate electrical and electronic devices which require to be completely sealed from the atmosphere. As the electrical resistance properties of the material are excellent, those components which do not require to be embedded in a block of resin for protection from mechanical shock, but merely need a thin protective skin, can be treated by dipping in a specially formulated liquid resin.

Epoxide resins have remarkable adhesive properties and special formulations are available for this purpose. A wide range of dissimilar materials can be joined together using these adhesives.

Laminated materials can be produced by impregnating glass fibres or glass cloth with the liquid resin. Laminated extrusions are produced by running glass fibres from their spools into a tank of resin and then through the die-head of an extrusion machine. Actuator arms and other components for use with electrical equipment are produced by this method.

The resins are tough and resistant to impact forces and some grades have extremely good dimensional stability, both during and after cure.

One particular disadvantage with epoxide resins is the high toxicity of some of the chemicals used in the hardeners. Care must be taken when handling these chemicals to ensure that contact with the skin is kept to an absolute minimum.

11.6 Mica

Mica is the name of a family of crystalline mineral substances found in large blocks between certain rock formations or as veins embedded in rock, dolomite, limestone, and other minerals. Owing to their appearance when extracted from the earth, large block deposits are called *books.*

The books are split into thin sheets which are processed into various shapes for use as electrical insulators; small pieces are collected and made into sheets by using plastics-based varnishes and adhesives, and mica dust is used as a filling material for plastics moulding powders and paints. Not all forms of mica are suitable for use as electrical insulators. The principal forms used are *Muscovite* mica and *Phlogopite* mica.

11.6.1 Muscovite mica

Muscovite mica, sometimes called *Ruby* or *Potash* mica, is almost transparent and has a glassy surface. It is coloured or spotted ruby, green, or brown. Clear ruby mica is of the best quality and is the hardest variety in common use. It is possible to split it into films 0·025 mm thick. Ruby mica is used in the manufacture of radio valves, electrical condensers, resistors, domesting heating and cooking equipment, cable joints, and where high temperatures up to 600°C are experienced.

11.6.2 Phlogopite mica

Phlogopite mica is also called *Amber* or *Magnesia* mica. It has pearl-like surfaces and is translucent. As one of its names implies, it is usually amber coloured, although it can also be reddish brown or green.

In general, it is a softer form of the mineral and is used for armature segment separators, commutator segment insulation, resistors, insulating washers and where temperatures up to 850°C are experienced.

11.6.3 Micanite

A trade name that has come to be commonly used to describe the pieces of thin mica film built up into insulation sheets, by using varnishes and plastics resins as adhesives, is Micanite. There are different methods of forming the sheets. One method is to immerse the tiny mica plates in a bath of varnish or resin and then press them between the platens of an hydraulic press which can be alternately heated and cooled. Another method is to mix the mica splittings with dry, powdered bonding material prior to the hot pressing stage.

Micanite is used as segment insulation in commutators, domestic heating appliances, slip-ring insulators, bus-bar insulation, rheostat resistance insulation, etc.

A moulding form of Micanite is available which has a higher plastics resin content than that supplied for the uses quoted above. The bonding resin can be either *shellac* or an *alkyd*, a material based on polyester resin. As supplied, the resin is only partly cured, complete cure taking place in the heated mould. On cooling in the mould the resin-bonded mica becomes hard and permanently set.

Cold-shaping flexible Micanite is made by using a thermoplastic bonding agent such as a silicone resin. This type of material may also be strengthened by having a base of glass cloth or paper.

Mica tapes are produced by spreading mica splittings on to a paper base and bonding with a thermoplastic resin. Mica-asbestos material is made by forming a sandwich of moulding grade Micanite between asbestos paper.

11.6.4 Processing mica

Processing mica requires special skills and techniques and it is advisable to employ the services of firms having the expertise in handling this brittle material rather than undertake forming operations which could damage both tools and mica.

Mica sheets are formed to shape by stamping, shearing, sanding, and cutting with a band saw. It can also be drilled. No lubricant or coolant can be used during these operations.

11.7 Vulcanized fibre

Vulcanized fibre is a hard, tough laminated material which is made by reacting good quality cellulose base material, such as paper or cotton, with sulphuric acid or zinc chloride. This process is followed by water washing and steam treatment and the resulting hydrated cellulose is pressed or formed into sheets, rods or tubes.

The material is usually coloured red, grey, or black and, although it is tough, it can be moulded to shape, drilled and tapped without difficulty. It is employed as a low-temperature insulation material in the form of base-boards, panels, stamped shapes and spacing washers.

Vulcanized fibre has a good resistance to abrasion and impact forces and is not attacked by petroleum products. Its specific gravity is about 1·4, which means that it is 40 per cent heavier than an equal volume of water.

11.8 Porcelain insulators

A typical electrical grade of porcelain consists of approximately 50 per cent clays, 25 per cent crushed and ground flint and 25 per cent crushed and ground feldspar. The clay is mixed with water and made into a *slurry* or *slip* and the pulverised flint and feldspar are added. The slip is processed through filter presses to reduce the water content and formed into clay cakes. The cakes are shaped by extrusion, moulding, casting, etc., and are finished by hand or by machining operations, depending upon the degree of hardness attained during the shaping and forming processes.

When the clay is dry, glaze is applied by spraying or dipping. Glaze is virtually clay slip to which has been added metallic oxides to give a colour, and feldspar and lime. The glaze cannot be applied to all surfaces since a free area must be provided so that the porcelain can rest on the floor of the kiln without the risk of fusion.

Immediately after glazing the porcelain enters the kiln for firing where it is *vitrified*, that is, fusion takes place throughout the mass of the material to form a glass bond that will make the porcelain impervious to liquids.

Insulator design must be carefully carried out because porcelain is used where the highest voltages are experienced in atmospheric conditions.

During manufacture the porcelain has passed through a high-temperature process which makes it resistant to the severe burning action during arcing. To reduce the possibility of continuous arcing, which can occur when the porcelain surfaces become contaminated, the distance between terminals is increased without increasing the over-all length of the insulator by incorporating in the design a corrugated outline. A typical example of this is shown in Fig. 11.1.

In addition to solid insulators, porcelain is used as an outer sleeve or bushing to carry insulated conductors through metal shells, floors, and walls. Large transformer insulators are typical examples. The porcelain bushing often contains a machined laminated material such as paper-based phenolic or oil-impregnated paper, with the conductor passing through the centre.

Porcelain is moulded into shapes suitable for use in connectors, fuse boxes, transformers, capacitors, switches and fuse holders.

In some high-temperature installations the heat resistance of porcelain combined with flexibility of the conductor is required. For such purposes interlocking porcelain beads are threaded over the conductor. Porcelain beading is often used on domestic appliances such as toasters and ovens.

11.9 Glass insulators

Glass is an inorganic material, that is, there is no carbon in its chemical structure. The principal raw material is silica which is pure sand. Some of the others are soda, calcium oxide, magnesium oxide, lead oxide, and boron oxide.

Other constituents are used to obtain particular properties such as colour, low thermal expansion, heat resistance, exceptional clarity, etc.

Glass manufacture commences with a thorough mixing of dry ingredients which have been reduced to small particles. A quantity of waste glass is added to the mix which is then melted in a furnace.

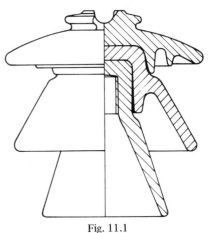

Fig. 11.1

A typical pin-type porcelain insulator

The furnace consists of a long, narrow structure made from refractory bricks and having an arched brick roof. The glass-making materials are charged into one end of the tank-shaped base of the furnace and a flame is played above the charge. Fusion and preliminary refining take place as the raw materials melt in the tremendous heat. The molten glass flows along the furnace, entering an area where the temperature is lower and full refining is achieved. A continuous flow of refined molten glass is now available for channelling to machines which form it to shape.

Electrical glass consists mainly of the following three grades;
(1)   *Soda-lime* glass, used to make electric light bulbs and other glass envelopes;
(2)   *Lead-alkali* glass, the principal material used for electrical insulators. A very high lead content produces a glass suitable for X-ray shielding;
(3)   *Borosilicate* glass, a high-quality electrical insulating material. This type of glass has a low thermal expansion and is produced for electrical and electronic devices which require a glass-to-metal seal.

11.10 Mineral insulation

The mineral referred to in mineral-insulated copper cable (M.I.C.C.) is magnesium oxide. The soft white insulation is highly compressed and packed around copper conductors and contained within a seamless copper tube. The whole assembly is then drawn to the required diameter.

Magnesium oxide insulated copper cables are fire-retarding and resistant to water, oil, and some chemicals. They do not deteriorate with age and, provided any cut surface is thoroughly dried and sealed, the cables will not absorb moisture from the air.

This type of insulation allows the cables to be bent to shape without difficulty.

11.11 Textile coverings

Cotton, silk and wool are frequently used as covering materials for flexible rubber and plastics insulators, and sometimes as insulators wound directly on the conductor. For certain applications they are impregnated with moisture repellants, and they can be coloured to assist in the identification of circuits.

11.12 Insulation failure

Disruptive breakdown occurs when there is a discharge between the conductor and the outer sheath. This can happen at a weak point in the insulation.

Under certain conditions of high electric loading the temperature of a dielectric will rapidly increase. This causes dielectric losses which raises the temperature still higher. When the generation of the heat becomes greater than the heat dissipation, the dielectric can catch fire with a consequent cable failure.

Tracking is the commonest form of failure of high voltage cables. This often starts at voids in the dielectric caused by the expansion and contraction that takes place as the electrical load increases and decreases. The impregnant deteriorates owing to ionization and tiny carbon particles are formed between the layers of insulating paper. This can build up to the point where a discharge path is formed and the insulation is destroyed.

Insulation failure will also occur when careless handling or faulty material causes severe mechanical damage to the cable.

11.13 Summary

The names of many common insulators and insulation systems are long and cumbersome to use, so they are known by their initials, and those mentioned in this chapter are as follows:

B.V.C.	black varnished cambric
Y.V.C.	yellow varnished cambric
T.R.S.	tough rubber sheathed
V.R.I.	vulcanized rubber insulation
GR-M or P.C.P.	polychloroprene (Neoprene)
GR-S	butadiene-styrene rubber (Government rubber styrene)
P.V.C.	polyvinyl chloride
P.V.-Ac	polyvinyl chloride-acetate
P.T.F.E.	polytetrafluoroethylene
P.C.T.F.E.	polytrifluorochloroethylene
P.F.	phenol formaldehyde
U.F.	urea formaldehyde
G.R.P.	glass reinforced plastics
M.I.C.C.	mineral insulated copper covered (cable)

The physical and mechanical properties are important when choosing a suitable insulator but the principal factors to consider are the working temperature and the time the insulator is exposed to that temperature. Increases in temperature can have the effect of drastically reducing the resistivity, especially when those increases are maintained for considerable periods of time.

Under normal working conditions an insulator may slowly deteriorate physically through ageing. This process is accompanied by a gradual loss of mechanical strength. When the material begins to disintegrate electrical failure will occur. Physical deterioration is accelerated at high temperatures and so selection of the correct material must take account of the working conditions. These influencing factors can result in the selection of a material having a higher electrical strength than is really required.

Insulators exposed to contaminants which can form flash-over tracks on their outer surfaces or cause the material to lose its desirable properties should be protected as much as possible before, during, and after installation to keep to a minimum the possibility of electrical breakdown.

Although the surface and volume resistivity of insulating materials are very important properties, the dielectric strength is often taken into account in the design of an insulator. Approximate dielectric constant values (relative permittivity) for a small number of insulating materials are given in the Table 11.1.

TABLE 11.1
Dielectric constants of some insulators

Material	Dielectric constant at 1000 Hz and 20°C
Glass	4
Mica	5
Insulating oil	2·5
Polythene	2·3
Bakelite	5
Porcelain	5·5
Urea formaldehyde	5
Polystyrene	2·5
Neoprene	6·5

12. TRANSFORMERS, RESISTORS, AND CAPACITORS

12.1 The transformer

The principal task of a transformer is to change the voltage of an alternating current from one value to another, either stepping it up or reducing it as required.

The essential parts of a transformer are two sets of windings of wire around a magnetic core, the wire itself being wound on formers made from suitable insulating material. The two sets of windings are called *primary* and *secondary* windings. When an alternating voltage is maintained across the primary winding an alternating flux is set up in the magnetic core. This induces an alternating e.m.f. to be set up in both the primary and secondary windings. Stepping the voltage up or down can be achieved by varying the number of turns of wire in the primary and secondary windings, because the voltages across them have the same ratio as the number of turns of wire in the windings.

12.1.1 Transformer cores

Transformer cores consist of thin wafers or laminations of magnetic material insulated from one another by varnishing or japanning one face of each lamination, with a minimum number of joints per lamination to maintain operating efficiency. The thickness of the laminations will vary, depending upon the circuit for which the transformer is designed. For example, for use at medium radio-frequencies each lamination could be as thin as 0·05 mm.

Insulated laminations are used instead of solid cores to reduce the harmful effects of the circulating eddy currents induced in the core material. This is done by preventing these currents passing from one lamination to another and restricting their flow to the high-resistance path offered by each individual lamination.

Core materials are based on ferrous metals, either silicon-iron, silicon-steel, nickel-iron or nickel-steel alloys.

Silicon-steel (see § 3.5) is produced in grades of varying permeability, that used in small power transformers and audio-frequency transformers having a relative permeability of about 8000. High permeability steel used in large power transformers has relative permeability values in excess of 25 000.

The cores in some transformers and induction coils used in telephony and radio design make use of what is called *directional materials* (see §16.2). These are usually silicon-irons or silicon-steels. The metal is annealed and cold rolled until the grain structure aligns itself with the direction of rolling. This type of metal has high permeability coupled with low eddy current loss, and when the flux path in the core is parallel to the direction of the grain structure, the transformer operates at its maximum efficiency.

Nickel-iron and nickel-steel alloys are produced specially for transformer construction. The principal nickel-iron alloys and their approximate respective relative permeabilities are:

Radiometal	20000
Rhometal	5000
Permalloy	80000
Mumetal	200000

Radiometal is used in both audio- and radio-frequency transformers, *Rhometal* is produced for low radio-frequency use, while *Permalloy* and *Mumetal* are used for small audio-frequency transformers.

Magnetic core material for use at high audio frequencies is sometimes ground to a fine powder, mixed with an insulating binding agent and pressed to the required shape. The materials used are powdered iron and a controlled mixture of iron oxide, zinc oxide, and manganese oxide.

12.1.2 Transformer windings

Transformer windings are produced from suitably insulated copper wire, the diameter of which will depend upon the current flowing and the heat generated.

The insulation can be either cotton, silk, plastics-based varnish, or enamel. The use of enamelled wire results in considerable space saving compared with wrapping-type insulation and allows the wire to be wound without cracking or peeling.

Immediately after winding, the coil should be impregnated with an approved wax or varnish. This assists protection against moisture, dirt, and mechanical damage.

12.1.3 Insulation

Transformer insulation falls into two main categories; low-temperature insulation such as paper, cotton, silk, varnish, enamel and P.F. laminates; and high-temperature insulation such as asbestos, mica, porcelain, and glass. The former group of insulators are cheaper than the latter.

Coil formers are made from P.F. laminates, mica, porcelain, and some plastics moulding materials.

12.1.4 Protection

Protection of transformers from mechanical and other damage is essential. In many instances the whole transformer assembly is enclosed in a steel casing and in large grid installations the casing would be filled with oil. The oil acts as a means of carrying away the heat generated in the transformer. It is pumped round the system and through a radiator which is exposed to a fan-driven flow of cooling air. The electrically-driven fans are automatically controlled by fluctuations in transformer temperature.

12.2 Resistors

The purpose of a resistor is to control the current flowing in any part of a circuit with a pre-determined degree of accuracy. In offering a resistance to current flow the resistor converts electrical energy into heat energy.

12.2.1 Wire-wound resistors

Wire-wound resistors are produced in a variety of different forms, some having exposed coils, while others are covered or sealed from the atmosphere. The wires used are made from nickel-silver, copper-nickel, nickel-chromium, copper-manganese, and iron-base alloys, all of which were described in Chapter 9.

A resistor wire should have a high resistivity, a melting point in excess of the maximum operating temperature and a low temperature coefficient of resistance to operate efficiently. Mechanically, the metal must be ductile, so that it can be wound on small diameter formers without cracking, it must be capable of being soldered or welded and the diameter of the wire must be held within reasonable limits of size.

The wires are wound on a variety of former materials including Micanite, insulated brass tubes, Tufnol and ceramics in various shapes. For radio-frequency use the material should have a low permittivity and the current flowing in adjacent turns of the winding should be in opposite directions in order to cancel the magnetic field and reduce the effect of self-inductance. This can be achieved by doubling the wire prior to winding, making two equivalent coils connected in series. The coils then carry current which flows in opposite directions.

Windings are often varnished or vitreous enamelled, or the wires can be silk-covered or enamelled before being wound. The method chosen will depend upon the rating of the resistor and the conditions under which it is expected to function. The resistance tolerance of most wire-wound resistors is ±2·5 per cent or less.

12.2.2 Composition resistors

Composition resistors are widely used in low-temperature telecommunications equipment, and consist of specially prepared resistance elements encased in moulded ceramic material or thermosetting plastics. A typical resistance element consists of a strictly controlled mixture of carbon black, talc and resin binder, pressed into shape and baked. Connections made from tinned copper wire are moulded in each end of the element and protrude through the outer covering which is moulded around the element. This type of resistor cannot achieve an accuracy better than ±5 per cent.

12.2.3 Carbon surface film resistors

Carbon surface film resistors are used at radio frequencies. The resistance element is made by baking a carbon film on a ceramic rod and then cutting a spiral groove in the carbon, leaving a helical strip of carbon of the required resistance adhering to the element support. End caps incorporating soldered terminals are sprung over the resistance element which is then completely sealed in a ceramic protecting case. The outer casing can be colour coded or the resistance value may be printed in figures. This type of resistor can be made to produce operating accuracies between ±0·1 per cent and ±2 per cent.

The advantages of both the composition and surface film resistors compared with wire-wound resistors are their high stability under storage or working conditions, low cost, low self-inductance, light weight and small size. The disadvantages are low operating temperatures, low power capacity of under 5 W, some internal noise generation, and the special care that must be taken when soldering into a circuit to ensure that heat is deflected away from the resistor to avoid permanent damage to the resistance element.

12.2.4 Variable resistors

One type of variable resistor consists of wire-windings of fixed resistance value wound on a former and held in a rigid frame; the variation in resistance being obtained by short circuiting or tapping off part of the total resistance value.

The resistance wire is usually a copper-nickel alloy and is wound on a porcelain or insulated steel former that may be hexagonal or circular in section. The windings terminate at nickel-plated brass sleeves.

The sliding contact is mounted on a nickel-plated brass rod of square or hexagonal section and holds a copper-graphite brush. Contact is maintained with the resistance wire by strong springs and wear is kept to a minimum by the self-lubricating properties of the graphite. The slider is made from a P.F. moulding powder ('Bakelite') and is of sufficient length to ensure that it does not bind on its support rod.

12.2.5 Colour coding

Colour coding is used to ensure that very small resistors can be identified (small figures can be difficult to read, and when surrounded by other equipment the figures may be hidden from sight). Even on larger resistors colour coding is preferable because figures become illegible after a period of time.

The standard colours used and their resistance values are given in Table 12.1, and some resistors are shown in Fig. 12.1.

TABLE 12.1
Colour code for resistance values

Colour	First figure	Second figure	Multiplying power
Gold			0·1
Black	0	0	None
Brown	1	1	10
Red	2	2	100
Orange	3	3	1000
Yellow	4	4	10000
Green	5	5	100000
Blue	6	6	1000000
Violet	7	7	
Grey	8	8	
White	9	9	

12.3 Capacitors

In its simplest form a capacitor is made from two conducting plates or electrodes separated from one another by an insulating medium called a *dielectric*. The dielectric may be air, solid insulating material, or a liquid, depending upon the duty of the capacitor.

When a voltage is applied to a circuit containing a capacitor, an extremely small *leakage current* passes between the plates through the dielectric. However, the vast majority of the energy is stored in the dielectric, so that in effect the current is broken at the capacitor. As the stored energy accumulates, a point is reached where the resistance of the dielectric ceases and a sudden increase occurs in the current passing through it; this point is called the *breakdown voltage* of the dielectric. The capacitor discharges and commences the cycle of events all over again.

Capacitors offer complete resistance to direct current but allow for the passage of alternating current. In telephone circuits this property is used to separate speech alternating current from signalling direct current.

12.3.1 Paper and foil capacitors

These capacitors are the simplest available and are produced in two forms. In one type, the electrodes are made from strips of aluminium foil separated by insulating paper. The foil and paper strips are rolled tightly together and formed flat or oval in section. The capacitor is then vacuum treated and impregnated with petroleum jelly or chlorinated naphthalene. Tinned copper strips are introduced between the strips of foil to act as end connexions.

In the other type, finely ground tin powder is pasted on one side of a roll of insulating paper. Two strips of this paper are rolled together so that in each roll the layers of tin are separated from one another by the paper which acts as the dielectric. End connexions are formed by inserting tinned copper strips between the layers to lie in contact with the very thin tin coating. The capacitor is vacuum treated and impregnated as previously described. This type of capacitor is called a *Mansbridge* capacitor, after its originator. Each of these capacitors is housed in a metal case.

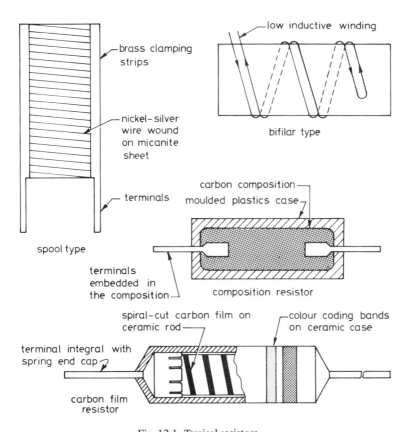

Fig. 12.1 Typical resistors

12.3.2 Other capacitors

Other capacitors used for particular purposes employ various materials as dielectrics. The principle of operation remains the same.

Thin mica plates silvered on both sides are used for some applications where paper will not operate efficiently. The mica plates are carefully stacked and riveted together, and the electrodes are connected electrically by means of silver-plated copper tags. Silver-plated copper end connexions are provided, held in place by the plate rivets. The whole assembly is vacuum treated, wax impregnated, and then sealed in wax or a plastics case.

Ceramic dielectrics are used in a number of different shapes to suit particular applications. In most cases the electrodes are formed by depositing a layer of silver on the two main surfaces of the ceramic dielectric. This type of construction is employed where very high electric strength, high permittivity, and controllable temperature-capacitance are required.

12.3.3 Electrolytic capacitors

Electrolytic capacitors are produced in two forms called wet and dry.

The wet type consists of an aluminium can holding an electrolytic solution of ammonium borate or sodium phosphate. The can forms the negative electrode (cathode), while the positive electrode (anode) is made from corrugated aluminium foil, held in place in the electrolyte by a stem which passes through a rubber gland. The gland also serves as an insulator between the two electrodes.

Once assembled, the electrodes are connected to a d.c. supply, with the positive pole connected to the anode and the negative pole connected to the can, and the resulting electrolytic action produces oxygen. This causes a very thin film of aluminium oxide to form on the large surface area of the corrugated anode. The oxide film is a dielectric and, owing to its large surface area, high capacitance is obtained from a small component.

In use the polarity described above must be observed or the oxide film will be destroyed. Other disadvantages are that the capacitor must be mounted vertically and that the electrolytic solution evaporates.

The dry-type capacitor overcomes these disadvantages because it is made from two very thin aluminium foil strips separated from one another by two layers of insulating paper saturated with an electrolytic paste of glycol and ammonium tetraborate. The strip which acts as the anode has a very thin oxide film on both sides, while the cathode strip has none. The foil and paper strips are rolled up and sealed at the ends with wax, and the whole assembly is contained in an aluminium or waxed cardboard container. Finally, Bakelite discs with connection tags fitted are used to seal the ends of the container.

Electrolytic capacitors are considered unsatisfactory for use in a.c. circuits because of the danger of experiencing the reversed polarity described above. Fig. 12.2 shows a small number of capacitors.

All capacitors operate on a factor of safety varying from 3 to 6, depending upon the materials used. This means that the working voltage is limited to values between one-third and one sixth of the maximum.

12.4 Summary

Transformers are used to change the voltage of an alternating current from one value to another. Increasing or reducing the voltage is achieved by altering the number of turns of wire in the primary and secondary windings. Insulated laminations are used in transformers to reduce eddy current losses to a minimum. Transformer cores are made from magnetic materials.

Resistors are used to control the current flowing in a circuit. Resistors can be either wire wound or made from suitable carbon material.

Capacitors use a dielectric material to store electrical energy for a given period of time, storage and discharge taking place automatically. Capacitors resist direct current but allow for the passage of alternating current.

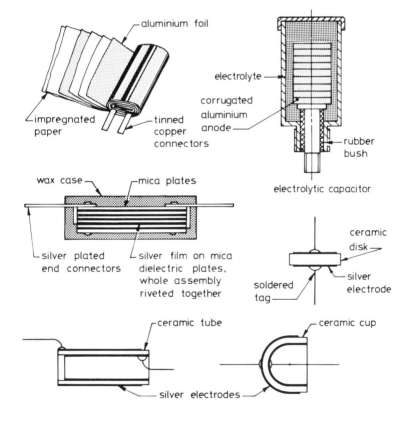

Fig. 12.2 A selection of capacitor designs

13. CABLES, LINES AND SUPPORTS

13.1 Underground electrical cables

These cables are either installed directly in the soil, in previously laid glazed earthenware ducts, or in troughs made from creosoted timber or earthenware which are filled with bitumen after the cables are laid.

When laid in open ground special care is taken to firm the earth around the cables, and interlocking tiles are laid in the soil about 75 mm above the cable to act as a warning to any future excavators.

The cables consist essentially of copper conductors covered with impregnated paper insulation and sheathed over-all with lead.

The conductors are made from high-purity electrolytically refined copper and are stranded to give the greatest possible degree of flexibility.

The insulation is generally wood-pulp paper applied in layers and impregnated with a mixture of heavy-grade mineral oil and resin. This offers resistance to moisture and also allows the layers of paper to slide over one another during handling.

The cables principally used for power transmission at high voltages are single cored or three cored. The single cored cable has the advantage of ease of construction and in a three-phase supply three similar cables are laid closely together, the lead sheaths being bonded together and earthed.

Three-core cables have the separate conductors housed inside a single lead sheath. The three cores are spirally wound together, the spaces between them being filled with jute or paper to produce a circular section. The cables are then covered with layers of insulating paper, which form what is called *belt* insulation.

The paper insulation is impregnated and the whole assembly is covered with a lead sheath which is then passed through a bath containing a bituminous compound.

Finally, the cable can be *armoured* by wrapping it with a layer of galvanized steel wire or layers of steel tape. This gives added protection from mechanical damage.

13.2 Underground telephone cables

These cables are more complex and more expensive than electrical cables because large numbers of wires must be formed into a core and encased in a single sheath. Coding systems are incorporated to enable each individual circuit to be identified.

The conductors are made from high-purity copper wire and are run as *singles, pairs, triples, quads,* etc., according to the duty of the circuit. To reduce interference in speech circuits to a minimum the wires are run as twisted pairs. The term pairs is used to describe two related insulated conductors forming a circuit, while quads is the term for a group of four insulated conductors. *Star-quad* cables are formed from a number of quads, each of which is made by twisting the four insulated conductors around a common axis.

The lay, or pitch, of a cable is the longitudinal distance taken for a quad or pair to make one complete helical twist.

Insulation materials are mainly paper, textiles, rubber P.V.C., and polythene. Paper insulation is wrapped around each conductor and given an uneven surface by crinkling it, so that a good volume of air is contained between each covered conductor. The air, being a dielectric, contributes to the over-all efficiency of the insulation.

Textile insulants include cotton, silk and wool. In many cases these can be colour coded to identify the circuits. P.V.C. is in widespread use both as an insulator and protective sheath. Identification of circuits covered with pigmented P.V.C. is achieved by coloured, spiralled markings.

Sheathing can be made from lead, lead alloys, P.V.C., or polythene. The important properties required are flexibility, resistance to corrosion and toughness.

A typical modern telephone cable consists of a large number of tinned copper conductors each covered with a sleeve of coloured P.V.C. The conductors are arranged in *pairs* or *triples* and formed into a core which is then covered with lappings of polythene-base tape. The whole assembly is sheathed with cream-coloured P.V.C.

In telephone exchanges the cable is made up of enamelled or tinned copper conductors insulated with P.V.C. and lapped with cellulose tape. The tape is coloured for identification purposes, and after lapping a layer of lacquer is applied to seal the joints. The wires are arranged in *quads* and formed into a core which is sheathed with cream-coloured P.V.C.

Where screening is necessary in telephone and transmission cables, helically wound aluminium tape is used to cover the core. For bonding purposes a tinned copper strip is positioned over the aluminium tape, running the length of the cable. The whole assembly is then sheathed with P.V.C.

13.3 Electrical transmission lines

All conductors used for overhead transmission of electrical power are stranded to give maximum flexibility. The usual method of construction is to wind helically successive layers of wire around one central wire, the consecutive layers being wound in opposite directions in order to hold all the wires together in what is termed *concentric lay*. Where large cross-sectional areas are concerned a *rope lay* is often used to obtain maximum flexibility.

The conductors are made from copper, aluminium, steel-cored aluminium, copper-clad steel, and sometimes steel alone. Copper is the most widely used metal but aluminium is becoming more popular owing to its lower cost and lightness. Even though the diameter of an aluminium conductor must be about 1·25 times that of a copper conductor of equal resistance, the mass of the aluminium is only one-half that of the copper. Aluminium conductors having a central core of galvanized steel wires are being used increasingly for long spans between towers of normal height.

Copper-clad steel wires have immense strength compared with pure copper or aluminium conductors. This type of wire is used for river crossings or where very long spans are necessary.

13.3.1 Mineral-insulated cable

Mineral-insulated copper cable (M.I.C.C.) consists of conductors encased in the mineral *magnesium oxide* and sheathed with copper (see § 11.10). The greatest care must be taken when preparing and completing the cable ends for terminal connection. No moisture must be allowed to come into contact with the insulation owing to its hydroscopic (water-absorbing) nature. For this reason special procedures and tools are required to complete a connection. The components for cable-end sealing should be prepared in advance of cutting the cable to minimise moisture absorption, and where this might occur a blowlamp can be used to dry out the dielectric prior to sealing. M.I.C.C. is fire-resistant and can be used in conditions where the temperature rises to 250°C. The insulation allows the cable to be bent to shape quite easily.

13.4 Telecommunications lines

Overhead lines may be bare or covered, although covered conductors are limited in use to positions where they pass through corrosive industrial atmospheres, where tree cutting is not permitted, over railways and bridges, and where underground cables would be too expensive.

Bare conductors are usually made from hard-drawn copper and cadmium-copper alloy. Overhead lines must be stronger than equivalent underground lines owing to the need to sustain their own mass, resist wind forces and the build-up of ice and snow.

Covered conductors are made from cadmium-copper alloy sheathed in black P.V.C. insulation. Drop wires to buildings are insulated with vulcanized rubber or P.V.C.

13.5 Line supports

Supporting structures for overhead electrical and telecommunications transmission are often similar. The exception is where long-distance National Grid lines are concerned; these require braced steel towers.

Wood poles are made from softwoods such as pine, fir and larch; the most suitable of these being the Scots pine. The advantages of using wood poles are their natural elasticity, relative cheapness, long life, the fact that no elaborate foundations are required, ease of erection without the need for special tools, and their comparative lightness. The disadvantages include long seasoning time (up to two years), the need to impregnate the wood with creosote or other preservative under pressure after seasoning, the possibility of fungal attack, weathering decay, and infestation by insects.

Testing of wood poles *in situ* consists of three methods; hammering, prodding and boring. Tapping the base of a pole with a light hammer gives a sound indication of the condition of the wood at the striking point. A hollow sound indicates severe decay and a dull sound indicates slight decay. Prodding is carried out using a sharp-pointed steel tool. Decayed wood will not resist penetration.

When the internal condition of a pole is in doubt a specially made boring bit fitted in a carpenter's brace is used. Decayed wood will offer little resistance and the existence of cavities is indicated by no resistance and the ability to push the bit backwards and forwards.

Reinforced-concrete poles are used occasionally. The poles are specially made by factory processes. Initial costs are low but the mass of each pole is excessive, and so transporting costs are high and erection is more difficult than with a wood pole. An attractive feature is the long, maintenance-free life of this material.

Steel towers are used for long-distance transmission of high-voltage current and are broad-based structures made from galvanized steel sections bolted together. Concrete foundations are essential and the tower sites require large areas of land.

13.6 Summary

Cables may be underground or overhead depending upon circumstances and cost. Electrical conductors are made from copper, aluminium, steel-cored aluminium or copper-clad steel. Telecommunications conductors are made from copper or copper-cadmium alloy.

Underground cables are fully insulated and sheathed, while the majority of overhead cables have bare conductors. Insulation materials used with underground cables include paper, textiles, P.V.C., and polythene. Oil and wax impregnants are used and the cables are sheathed with lead or P.V.C. Telecommunications cables have each of the large number of conductors covered with a colour-coded insulator for ease of identification of the circuits.

Wood poles supporting overhead lines are made from seasoned softwoods impregnated with creosote under pressure. Steel poles and towers are made from galvanized steel.

SECTION II
PROCESSES

14. FOUNDRY PROCESSES

14.1 Metal casting

In its simplest form a mould suitable for casting shapes in metal can be made by pressing a solid pattern of the article to be cast into specially prepared sand. The removal of the pattern leaves an impression in the sand into which molten metal is poured. When the metal has solidified and cooled it is removed from the mould so that further work can be carried out to produce the finished article.

In practice, the processes and the materials used in foundry work are carefully controlled, and a high degree of technical knowledge and skill are required to produce good-quality castings.

14.1.1 Moulding sands

Moulding sands are specially selected, washed and blended to make them suitable for mould-making. Mechanical crushers are often used to adjust the particle size of the sand and to improve particular characteristics.

The sands must have *porosity* to allow air and moisture to escape and ensure the liberation of gases generated in the mould when the molten metal is poured. *Plasticity* is essential to enable the shape of the pattern to be faithfully followed by the sand. As the sand must cling to the sides of the container or *flask* within which the moulding takes place, *adhesion* is necessary. The particles of sand must also stick to one another with ease and so *cohesion* is required; this characteristic also ensures that the sand will not break away when the pattern is removed or when the molten metal flows over it. The sand must have a refractory nature since the fusion of sand particles with the surface of the casting would cause machining problems at later stages of production.

14.1.2 Sand moulds

Sand moulds can be used while moist, in which case the name *green-sand mould* is used to describe the moulding conditions. The alternative to this is *dry-sand moulding* when the mould is dried in an oven before the metal is poured. (Other methods of treating sand, such as the use of plastics resins and carbon dioxide, are employed where high productivity of repetitive parts is concerned, but these processes are outside the scope of this book).

In green-sand moulds substances such as charcoal, powdered coke, and coal dust are blended with the sand. Apart from colouring the sand black these materials absorb some of the heat generated within the mould and also improve the porosity.

In dry-sand moulds heavier grades of close-textured sand are used; horse manure, fine straw, or animal hairs are mixed with the sand to allow the same degree of venting as in green-sand moulds.

The sand faces in contact with the metal are often painted or dusted with substances having a high refractory nature to prevent sand burns appearing on the castings. The best material for this purpose is a form of carbon called *blackings*.

14.1.3 The moulding flask

The moulding flask containing the sand mould can be made from wood or metal. It consists of two parts, the *cope* at the top and the *drag* at the bottom. The two frames are machined across the joint face to ensure a snug fit and dowel pins are provided to locate the two parts when the mould is closed. Before pouring metal into the mould, cotter pins are fitted in the dowels to prevent the cope and drag separating.

Bars running across the frames, or grooves machined into the sides, are sometimes provided in each half of the flask, principally to give additional area to which the sand can adhere. The bars often give additional rigidity.

Two halves of a typical moulding flask are shown in Fig. 14.1.

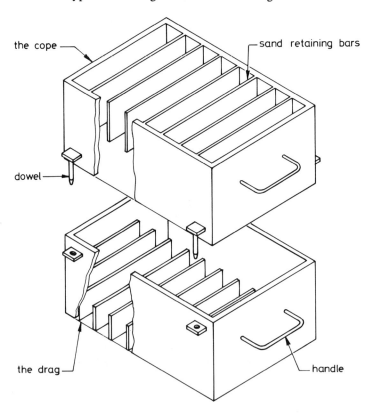

Fig. 14.1 A simple two-part moulding flask

14.1.4 Patterns

The patterns are usually made from wood and painted to give a smooth finish. A high degree of skill and a sound knowledge of foundry work are required of the pattern-maker, since he will decide how the casting can best be achieved, where the moulding joints are to be, and so on. Since the pattern must come away from the sand easily and without causing damage, any vertical faces are made to taper inwards towards the bottom of the impression.

As metal cools in the mould it contracts, and to account for this the pattern is made oversize, the difference in dimensions depending upon the type of metal used. The pattern-maker has a special rule, the unit divisions of which are slightly longer than normal so that he can work directly from a drawing without having to calculate allowances for contraction in the moulding.

A typical pattern is shown in Fig. 14.2.

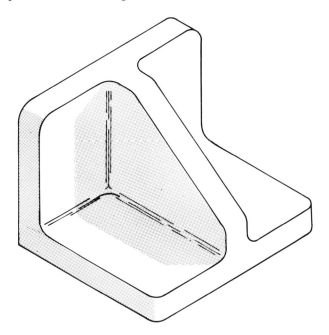

Fig. 14.2 Pattern for a cast iron bracket

14.1.5 Producing a mould

The method of producing a mould for a cast-iron bracket from the pattern shown in Fig. 14.2 will be described up to the point just prior to pouring the metal. The series of diagrams in Fig. 14.3 indicate the sequence of operations.

(1)   The drag is placed joint-face downwards on a *moulding board* or *turnover board,* the pattern is correctly positioned, and facing sand followed by moulding sand is rammed around it. It is necessary to ram the sand to ensure that the moulding cavity does not enlarge when molten metal enters, but it must not be compacted too much otherwise the air and gases cannot escape.

(2)   Another board, called a *bottom board*, is placed on top of the drag and the whole assembly is turned over so that the joint face is uppermost. The turnover board is removed and the facing sand is carefully inspected to see that it is flush with the joint and well compacted.

(3)   The cope is positioned on the drag, and tapered wooden or metal pegs are placed in the positions where the molten metal will be poured and also where it will rise when the mould cavity is full. Moulding sand is placed in the cope and around

the pegs and is rammed into place. The tapered holes left in the sand when the pegs are removed are called the *runner* and *riser*.

(4)   The cope is removed and the joint face inspected for any imperfections. A cavity or small depression is cut into the sand in the drag immediately beneath the position of the runner, using special tools. From this depression a channel or *gate* is cut leading to the mould cavity. The pattern is removed and the mould faces in contact with metal are painted with a refractory compound.

(5)   The cope and drag are reassembled and prepared to receive the molten cast iron.

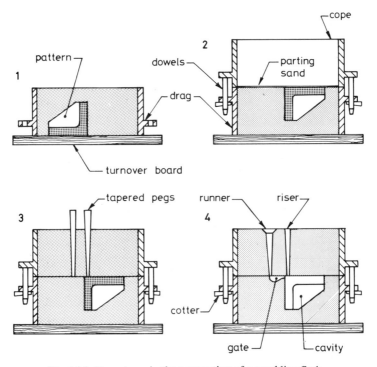

Fig. 14.3 Four stages in the preparation of a moulding flask

The moulding will not be removed from the mould until it has settled and cooled. Upon removal, the bracket will have attached to it the runner and riser and excess metal called *fins* which form along the parting line between the cope and the drag. These are removed and the casting is given a general cleaning. This series of operations is called *fettling*.

When smooth flat faces are required on the flanges, or when holes are called for, the casting is machined.

Cores are introduced into a mould to take up space and form hollow sections or holes and also to keep to a minimum the amount of metal used and machining time.

Cores are moulded from sand in small wooden moulds specially made for the purpose. The sand is mixed with special compounds called *binders* so that the cores retain their shape. It is often necessary to use reinforcing material such as wire or metal rods to enable the cores to withstand the forces imposed by the molten metal. Long or heavy cores are supported either in *core prints*, which are depressions left in the mould-ing sand by material specially added to the pattern, or by devices called *chaplets*. All of these items are depicted in Fig. 14.4.

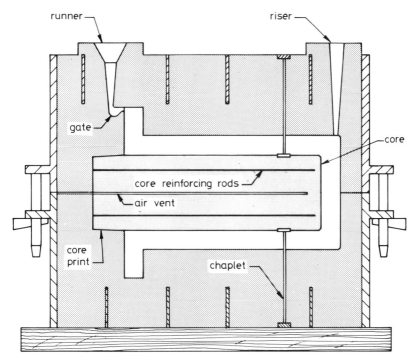

Fig. 14.4 A mould showing a core print and chaplets

14.1.6 Casting metals

Casting metals include cast iron, steel, aluminium and copper alloys such as phos-phor bronze and brass. All these metals, except cast iron, can be supplied in forms other than as castings.

Casting is a cheap way of producing metal parts of complicated shape in large num-bers, provided that the metal itself is cheap and that the casting processes are more economical than machining or fabrication methods.

High melting-point materials such as steel and cast iron are moulded in sand, while the lower melting point metals such as aluminium-zinc alloys, typemetal and lead can be moulded in permanent steel moulds.

Sand moulds are broken open to remove the finished castings, with a consequent waste of material and a long casting cycle. Permanent steel moulds, which have a high initial cost, can be used many times over, have a quicker casting cycle, and are more economical in the use of the casting metal. The permanent steel moulds are called *dies* and the casting process itself is called *die casting*.

14.2 Summary

Mould cavities suitable for casting are formed by pressing a solid pattern in specially prepared sand. Moulding sands are specially selected and treated before use. Sand moulds can be used in the moist or dry condition. This gives rise to the terms green-sand moulding and dry-sand moulding. A moulding flask consists of two principal parts, the cope at the top and the drag at the bottom.

Patterns are usually made from wood and painted to a smooth finish. Metal is poured into the runner and appears at the riser when the mould is full. Mouldings have excess metal removed during fettling operations. Some metals are suitable for die casting using steel moulds or dies.

15. FORGING

15.1 Forging principles

Forging is a process of shaping metal by first heating it in an open hearth or a furnace until it is red-hot and then manipulating it into shape by subjecting it to hammer blows or the squeezing force of powered hammers.

It is often necessary to return the work to the furnace to maintain the metal in a state suitable for shaping. The traditional hand work of a blacksmith is forging. He uses a number of different tools to obtain specific results but his basic equipment is a hearth to heat the metal, an anvil on which to shape it, and a hammer to deliver the force required to make the shape. Some of the basic tools are drawn in Fig. 15.1.

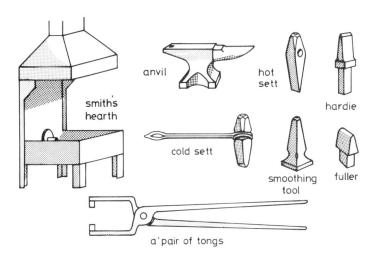

Fig. 15.1 A small number of blacksmith's tools

The modern blacksmith is very skilled in the use of hand tools and power hammers, and this skill is required when presenting hot metal to the hammers to produce the best possible product.

Hand skills are confined to the manufacture of 'one off' jobs and small, specialized components. The production of large numbers of identical parts is carried out by using steel dies and powered hammers.

Forged parts produced in steel dies waste little or no material, have a good finish, and can be made to high standards of accuracy. As the metal is punched or pressed into a die it flows to take up the desired shape. This means that the grain flow follows the shape of the die and is continuous. When a similar part is machined the grain flow is interrupted where the cutting tool removes metal. The result is that the forged part has greater strength than the machined part, provided that the metal is the same. An example of grain flow is shown in Fig. 15.2.

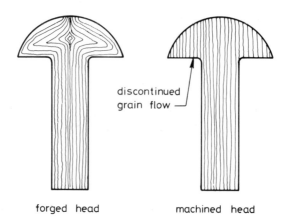

forged head   machined head

Fig. 15.2 Grain flow in a forged part

15.1.1 Upsetting

Upsetting is a method of producing rivets, nails, screws and bolts, the heads of which are formed by hammering the locally heated ends of pieces of rod or wire. The principles involved are shown in Fig. 15.3.

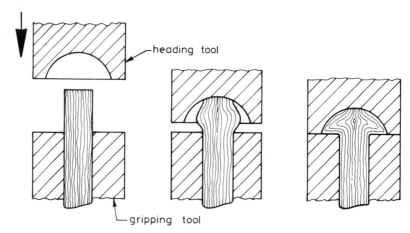

Fig. 15.3 The principles of upset forging

The rod to be upset is heated at one end then gripped in a die block. The hammer descends and forces the head into shape with a heavy blow. There is no waste material and the grain flow of the metal will be continuous.

15.1.2 Drawing and setting down

Drawing and setting down are processes undertaken by the blacksmith using anvil and hammer.

When it is necessary to lengthen a piece of metal it is *drawn down*. This process is carried out using the rear edge of the anvil and a hammer, but for large pieces of metal a powered hammer is used. The heated material is struck a series of blows along its length which makes it spread outwards. There is an increase in length, a slight increase in width, and a change in the section of the metal, which becomes progressively thinner. The tools used are called *fullers* when the work is of square or rectangular section and *swages* when the work is circular in section.

Drawing down leaves the surfaces of the work in a ridged state, therefore a follow-up process is necessary to smooth out the ridges and give a good finish. This process is called *setting down* or *flatting*. The processes mentioned in this section are described diagrammatically in Fig. 15.4.

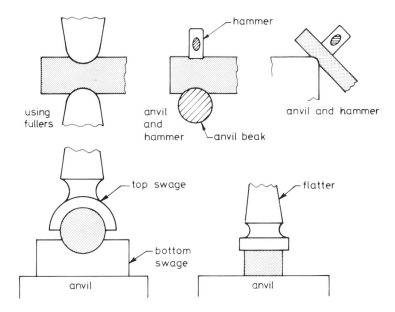

Fig. 15.4 Drawing and setting-down processes

15.1.3 Bending

Bending utilizes the face, sides, or beak of an anvil. Small pieces can be hammered over the edges of dies held firmly in the jaws of a bench vice, when the bending must take place away from the operator and towards the fixed jaw of the vice. Hammering towards the movable jaw could damage the vice mechanism. Die pieces should be used in vice work for two main reasons; to prevent the serrated faces of the vice jaws damaging the work and to obtain the correct radius and angle of the bend.

The diagrams in Fig. 15.5 describe these operations.

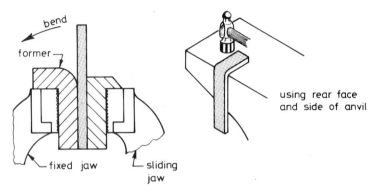

Fig. 15.5 Simple bending processes

15.2 Summary

Forging is carried out on red-hot metal parts to fashion them to a required shape and size.

The processes may be executed entirely by hand or by using power presses and dies.

Forged and upset metal has a superior grain flow compared with machined metal and is therefore stronger.

16. ROLLING

16.1 Hot rolling

The ingot is the first stage in the production of steel after the metal leaves the furnace. The ingot is a large steel casting having a mass between 2000 kg and 10000 kg. Before it is finally processed the ingot must be reduced in size and shape. The initial reduction is made while the steel is white-hot either in a *cogging mill*, which produces blooms, or a *slabbing mill*, which produces slabs, (see Fig. 3.3). The mills are massive structures which support extremely hard and heavy rolls driven by powerful electric motors. There are usually only two rolls in these primary processing mills mounted one above the other and known as *two-high mills*. The direction of rotation of the rolls is reversible enabling the white hot metal to be passed backwards and forwards between them. Built into the mills' structure are heavy manipulating plates that can turn the ingot over, straighten it, or move it from side to side on the mill bed.

As the ingot passes through the rolls it is reduced in section and increased in length; the distance between the rolls is reduced after each pass. When two passes have been completed the ingot will be reduced in thickness by about 100 mm; it is then turned over and passed through twice more making the section approximately square. The process of rolling and turning continues until the correct size of bloom or slab is reached. Alternatively, the reduced ingot can be passed on a roller-conveyor through further rolling processes while it is still sufficiently hot to be manipulated. These mills increase in processing speed as the ingot becomes thinner and longer, eventually making the ingot into sheets or strip suitable for the manufacture of a wide variety of pressed shapes. The process is often continuous and, during its travels through one set of rolls after another, the metal becomes quite cold. This means that, finally, *cold rolling* takes place and this gives the steel its polished finish. Fig. 16.1 shows a diagrammatic representation of a two-high mill rolling a slab.

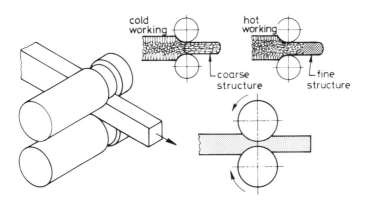

Fig. 16.1 A two-high rolling mill

Hot rolling produces *black mild steel* because the hot steel looks black, a feature caused by rolling into the surface of the metal the oxidized scale which rapidly forms on the outer surfaces. The scaly surface acts as a barrier to the elements and delays the onset of corrosion. Heavy plates, steel girders, angles, and channel sections are hot rolled.

Fig. 16.2 shows a diagram of a steel channel passing through finishing rolls. The bloom requires a number of passes through rolls of different shapes until it reaches this stage and these different sections up to the completed girder are also shown.

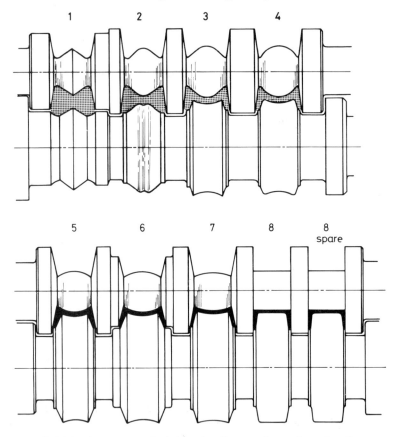

Fig. 16.2 The passes required through rolls to produce a channel section

16.2 Cold rolling

Hot-rolled steel that is to be made into sheet or strip must first have the scale removed from its surfaces. This can be achieved either by pickling in an acid bath or by cracking the iron oxide into small pieces between the rolls and removing them by high-pressure water jets. When the steel is pickled it must be washed and dried before it passes to the cold-rolling mills. Cold-rolling mills are usually four-high, the smaller finishing rolls being backed up by heavy rolls to withstand the tremendous forces. The rolls are highly polished and this gives the steel a bright, smooth finish. The steel which passes to the cold-rolling section is first of all reduced almost to size by hot rolling. In addition to having a better finish than black mild steel, cold-rolled steel is also harder and stronger owing to the work-hardening effects of the process.

Large reductions in size are not possible by cold rolling because the steel is no longer is a plastic state, but the grain structure of the metal is influenced sufficiently to align itself in the direction of rolling.

This *directional material* as it is called can cause manufacturing difficulties and it must be treated in somewhat the same manner as grained timber. This property of cold-rolled steel can be offset by rolling the metal in a number of different directions but this makes the steel more expensive. Directional steels and other alloys are used to make cores for transformers and induction coils as mentioned in § 12.1.1.

The product of the cold-rolling mills is known as *bright mild steel*. It is more accurately finished to size than black mild steel but the corrosion resistance of its shiny surfaces is nil. For this reason bright steel products are oiled or greased immediately after manufacture to exclude the air.

Diagrams of two-high, three-high, and four-high cold-rolling stands are shown in Fig. 16.3.

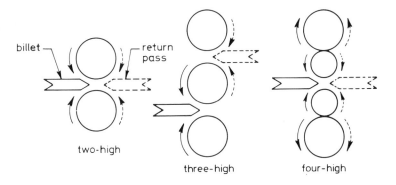

Fig. 16.3 Two-high, three-high and four-high mills

16.3 Annealing

Cold rolling hardens the metal, so it must be *annealed* before being passed on to the customer for further processing. To exclude oxygen which would form a scale and ruin the surfaces, the steel is heated in special containers from which all the air is pumped out and replaced by an inert gas. The annealing furnace heats the steel to about $700\,^\circ$C and the metal is not removed until it has cooled in the furnace. This process ensures that the steel remains with a bright finish and is called *bright annealing*. The oil or grease protection mentioned earlier is applied after the bright annealing stage.

16.4 Summary

Steel ingots are hot rolled to blooms and slabs from which black plate, rails, channels, joists, etc. are made. Cold rolling gives previously hot-rolled metal a bright, shiny finish. Cold rolling can produce material whose grain structure is aligned with the direction of rolling. This directional material is used for a number of electrical applications. Cold-rolled metal is bright-annealed before delivery to customers for further processing. The processes described in this chapter are also applicable to non-ferrous metals.

17. DRAWING AND EXTRUSION

17.1 Wire drawing

Ductile metals such as mild steel, copper, and brass are made into rod or wire by the *drawing* process, which consists of pulling the metal through a series of dies of gradually decreasing diameter. As a rod is drawn through a number of dies to make it into wire of very small diameter, there is a considerable increase in length, so that a single copper wire-bar of mass 127 kg can provide approximately 100 km of wire 0·4 mm in diameter.

Although the method of drawing is basic to all metals, there are certain small differences in technique adopted for each one, therefore the processes described here will apply to copper and copper alloys.

17.1.1 Copper wire

Copper wire is produced from wire-bars which are heated in a furnace until they are red-hot. Each one is rolled into rod of about 8 mm diameter and is supplied to the wire manufacturer in coils.

A single coil of rod will not provide the amount of wire needed to make a large reel of stranded copper wire or cable and so a number of coils are butt-welded together to ensure that the wire-drawing machines can operate continuously.

17.1.2 Drawing machines

Wire-drawing machines have a number of die-heads, and the dies themselves are made from tungsten carbide for general work, and diamond for the very fine wires of 1 mm diameter and below. Tungsten carbide is an extremely hard material that is also used to make tips for metal-cutting tools. A schematic drawing of a multi-die wire-drawing machine and a typical die design are shown in Fig. 17.1.

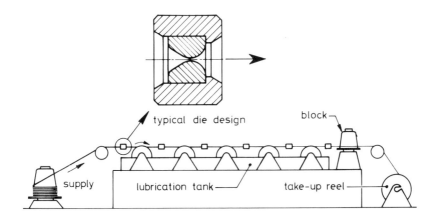

Fig. 17.1 Schematic layout of a multi-die wire-drawing machine

17.1.3 Seamless tubing

Seamless tube production starts with the casting of a cylindrical billet. This is made into a *tube shell* by forcing the red-hot billet over a mandrel of the required hole size and through a die-head which determines its external diameter. The tremendous forces necessary for this operation are provided by hydraulic rams which push the billet through the die in much the same way as toothpaste is forced through the nozzle of its tube when pressure is applied. This process is called *extrusion*.

The tube shell is worked at one end to provide a gripping point and is then fixed into a drawing machine which pulls the cold tube through a sizing die and over a sizing plug. The plug can be attached to a long rod fixed to the draw-bench, or it can be a floating plug which is kept in place by the action of the tube being drawn through the die. Machines fitted with this device can draw longer lengths of tube than those having a fixed plug. Lengths of straight tube which can be supplied are limited to about 20 m, as this is the maximum length available in transporting vehicles. Some tubing is supplied coiled in longer lengths and is straightened before it can be used.

Fig. 17.2 shows, in sketch form, the sequence of operations in the production of seamless copper tube.

17.2 Conduit

Conduit is made from strip metal that is folded into a circular or oval section by drawing it through a device called a bell. This is a die in the shape of a conical funnel, a sketch of which is shown in Fig. 17.3. The strip metal is usually supplied coiled on a drum and unwinds as it is pulled through the bell.

17.3 Extrusion

As mentioned above, the extrusion process entails pushing hot metal through a die under the action of hydraulic rams. The metal is easily worked while very hot and so a large variety of sections are made by extrusion, especially from the more ductile materials such as copper and its alloys, lead, and aluminium. Although a reasonable degree of dimensional accuracy is obtained in extruded products, it is often necessary to cold draw the extruded section when a tight control over dimensions and an increase in strength are required.

Cable coverings of all types are produced in a form of extrusion press. The insulated cable is fed through the centre of the machine and as it passes through the die the metal covering is extruded around it. The rate of feed of the cable must match the rate of extrusion.

Fig. 17.4 shows diagrammatically some methods of extrusion.

17.3.1 Impact extrusion

Impact extrusion is used to make collapsible tubes, dry-cell battery cases, transformer covers and other canister shapes used in electrical and telecommunications systems. The metals extruded include aluminium, lead, tin, zinc, and their alloys; the process being carried out at room temperature.

A comparatively thick disc of metal is located in a die, the diameter of the disc being approximately equal to the outside diameter of the finished canister. A punch

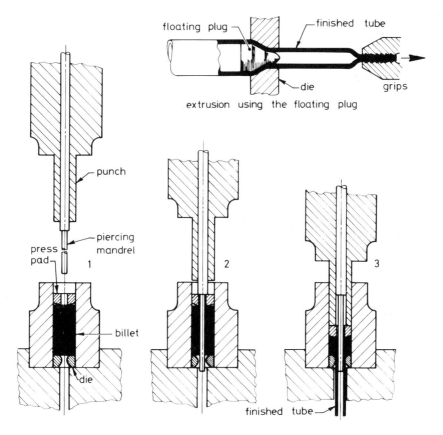

Fig. 17.2 The production of seamless copper tube

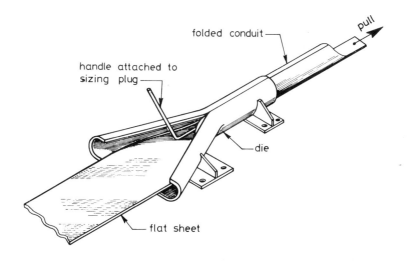

Fig. 17.3 Forming conduit by folding

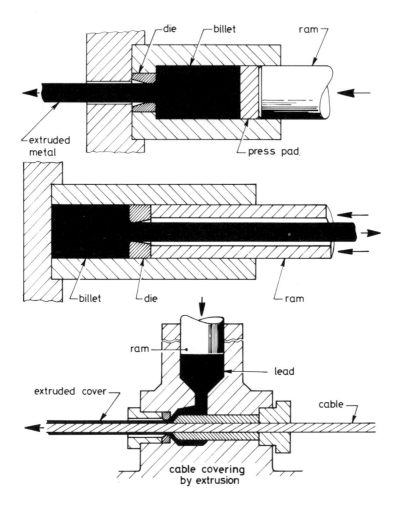

Fig. 17.4 Some methods of extrusion

descends swiftly with sufficient impact to extrude the metal between the punch and the die, forcing it to flow upwards and fold itself around the punch. The shape of the punch forms the inside shape of the canister which can have a conical, flat, curved, or bevelled bottom.

As the punch retracts from the die, a stripper plate removes the canister. The operation is very rapid, making possible high rates of production. The tools used in the impact extrusion process are often cheaper to produce than those used in other cold-extrusion techniques.

The extruded metals are in their fully work-hardened condition owing to the severe cold working process through which they pass.

Fig. 17.5 shows the impact extrusion process.

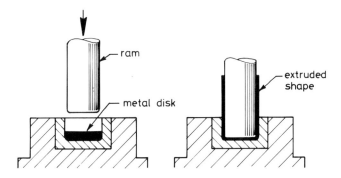

Fig. 17.5 Impact extrusion

17.4 Summary

Drawing is a method of pulling metal through a die or dies of the required shape and size. Extrusion involves pushing hot material through a die. Impact extrusion is a cold forming process by which metal is induced to flow around a swiftly descending punch. Drawing is usually confined to the production of circular sectioned articles such as wire, while extrusion can be used to produce a variety of sections.

18. PRESSING AND PRESS TOOLS

18.1 Processes

To produce large numbers of repetitive articles from sheet metal by blanking, piercing and pressing, specially made *press tools* are necessary. The forces required to shape the metal are supplied by some form of press, the type and size of which will depend upon the metal used and its thickness. Some presses are designed to deliver a heavy blow to the workpiece, while others are arranged to provide a steady squeezing action.

Blanking operations are carried out to provide a piece of metal cut to size, from which the finished article will be made. Blanking tools must be accurately produced from hardened alloy steel and the cutting edges must be sharp and tough.

Sometimes blanking and other operations are completed in a single tool, using only one stroke of the press. This keeps down the cost of production and often ensures a steady flow of finished articles which require little or no additional attention.

Piercing refers to the process of making holes of any shape in a blank, or in a sheet or strip, prior to blanking. The tools required for piercing must be tough and sharp and the punch and die are made to the size of the hole, the piece punched out being scrap.

Bending is the term applied to the production of one or more simple bends in sheet metal. A bending tool is shown in Fig. 18.1.

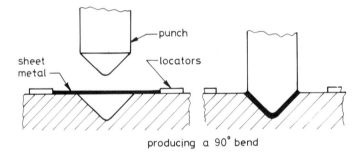

producing a 90° bend

Fig. 18.1 A simple bending tool

Forming or *pressing* generally refers to complicated bending processes such as recessing, indenting, and the formation of channel sections, etc., in sheet metal. A simple forming tool is shown in Fig. 18.2.

Drawing of sheet metal applies to the process of making cup-shaped articles in one piece. Deep drawing refers to drawn articles having a length far in excess of the diameter, e.g. cartridge cases. It is necessary to produce deep-drawn items in stages, each stage requiring a different tool. (Compare the meaning of *drawing* applied to sheet metal, and to wire as described in the previous chapter).

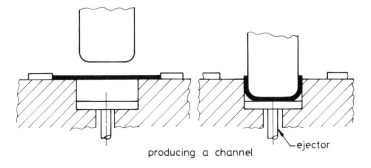

Fig. 18.2 A simple forming tool

18.2 Materials

The softer, more ductile materials are selected for pressing and stamping operations and even these are often annealed between operations so that cracks and severe wrinkling are avoided. The metals most suited to press work are copper, aluminium and its alloys, brass, and steel containing no more than 0·1 per cent carbon.

18.3 Summary

For the high-speed production of pressed shapes from sheet metal, press tools and powered presses are required. The principal operations are blanking out the developed shape; piercing and punching any holes; bending or forming; and drawing. Soft ductile materials are selected for this class of work. Tools must be tough, hard, and sharp.

19. SOLDERS AND SOLDERING

All solders are alloys; the so-called *soft solders* consisting mainly of lead and tin. The proportions of these two elements, and other small additions varying with the metals to be joined.

The *silver solders* are alloys of copper, zinc, and silver and are described in the section on brazing (in Chapter 20) because the alloys themselves and the techniques used in joining metals are more closely related to brazing than soldering.

19.1 Lead

Lead is an important constituent of solders and many other useful alloys. Pure lead is a very soft, bluish-grey metal having a high specific gravity of 11·3. Although mechanically weak, its ease of working and corrosion resistance make it a suitable material for cable sheathing. The plates in storage batteries are made from an alloy of lead and about 10 per cent antimony.

At 20°C the electrical resistivity of lead is 0·21 $\mu\Omega$ m and its melting temperature is 327°C. It is used as an alloying element on the metals from which a wide variety of useful articles are made including dry-cell battery cases, master patterns of complicated articles, small die-castings, filler material to aid tube bending, squeezable tubes, etc.

Fig. 17.4 shows, diagrammatically, how electric cables can be sheathed with lead.

19.2 Tin

Tin is a soft, silvery-white metal having a specific gravity of 7·3 and a melting point of 232°C. It is a very expensive but mechanically weak metal, making it unsuitable for use other than as an alloying element or as a thin coating on mild steel and other sheet material. Its electrical conductivity is about one-sixth that of pure copper and it is virtually non-toxic.

19.3 Soft solders

The melting point of soft solder varies according to the content of the alloying elements but it must always be lower than that of the parts to be joined and higher than the temperature expected to be reached in service.

Essentially, soft soldering means the introduction of a thin film of alloy between the mating faces of the parts being joined. In order to produce a sound joint the solder must flow freely over, and remain in contact with, the metals being joined, i.e., it must *wet* the surfaces. To achieve this the metal parts must be clean and free from grease or oxide layers.

Changes in the temperature range over which a solder remains plastic can be controlled by altering the proportions of lead and tin in the alloy. Because of this difference in behaviour the various grades of solder are used for specific purposes.

Eutectic solder has no plastic range at all; it melts and hardens at 183°C. The name *eutectic* signifies that the alloy solidifies at a constant temperature which is lower than the melting point of either of the metals in the alloy. This is also called *Tinman's solder*.

Wire solder used for soldering wires to tags, etc. contains 60 per cent tin and 40 per cent lead and has a small plastic range. Speed of solidification means economy in the soldering operations and a reduction in worker fatigue. This is sometimes called *Tinman's coarse solder*.

Cable sheathing often requires lead sleeves and these are soldered using an alloy containing 50 per cent tin and 50 per cent lead.

Plumbing solder is used where a greater plastic range is necessary so that joints can be wiped to shape. The alloy contains 32 per cent tin, 66 per cent lead and 2 per cent antimony. The addition of antimony is made to give the solder improved mechanical strength.

19.4 Fluxes

To achieve the correct amount of surface contact and have reasonable mechanical strength, the faces being joined must be perfectly clean and free from oxide film and grease. For these reasons fluxes are used which may be either *chemically active* or *protective*.

19.4.1 Chemically-active fluxes

Chemically-active fluxes are the so-called *killed spirits*, zinc chloride, ammonium chloride, and zinc ammonium chloride. These prevent oxidation of the joint faces and also partially clean the metals. Killed spirit fluxes are not frequently used in radio and telecommunications work because fusing of the salts does not occur below 262°C, a temperature much higher than the melting-point of the solder. This means that there is a possibility of unfused salts becoming trapped within the molten solder, resulting in brittle joints. The salts are also hydroscopic (absorb moisture), and moisture mixed with the tiny crystals forms a corrosive substance which can result in eventual failure.

19.4.2 Protective fluxes

Protective fluxes are made from resin, tallow, olive oil, petroleum jelly, and mixtures of these substances. No cleaning of the joint faces takes place but there is complete protection from oxide formation while the metals are raised to soldering temperature.

The flux used for telecommunications work is processed resin. The processing removes the non-volatile oils which would otherwise remain as a sticky coating on the joint faces. The resin is extruded with the solder during manufacture to form multi-resin cores that ensure a continuous supply of flux to the joint areas.

Table 19.1 contains a small selection of common solders giving percentage composition of the alloys and their applications.

TABLE 19.1

A small selection of soft solders

Composition (%)				Approximate solidification temperature range (°C)	Applications
Sn	Pb	Sb	Ag		
62	38	–	–	Exactly 183	Tinman's solder. Used for electrical, radio, and instrument work
50	50	–	–	220 - 185	Tinman's coarse solder. Used where a slight plastic range can be tolerated
32	66	2	–	238 - 190	Plumber's solder. Used for wiped joints in pipes and cables
43	55	2	–	225 - 185	A general-purpose solder
12	80	8	–	250 - 240	Higher temperature solder used for soldering steel articles
–	97	–	3	308	High temperature solder used for soldering copper and copper alloys

19.5 Soldering

The metals that can be soldered are iron, steel, copper and copper alloys, nickel, tin, zinc, lead, and aluminium and its alloys. Many non-metallic materials can be joined together by soldering provided that metallic surfaces are either sprayed or electro-deposited on the joint faces.

Before soldering can commence the articles to be joined must be thoroughly cleaned across their contact faces to ensure that no dirt, grease, oxide film or other impurity is present. Fig. 19.1 shows the preparations required prior to soldering.

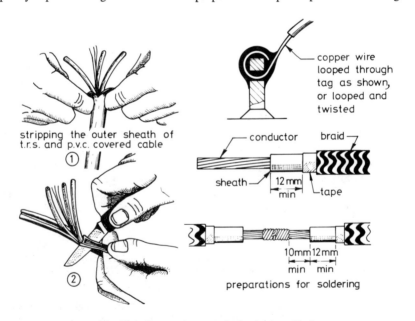

Fig. 19.1 Preparations required prior to soldering

The point of the copper *bit* of the correct size of soldering iron must be *tinned*, that is, given a thin coating of solder. Unless this is done the solder will not be correctly transferred to the joint. An efficient method of achieving this is to press a cleaned and heated bit into a hollowed block of ammonium chloride and then allow a few drops of molten solder to enter the holloe and come into contact with the bit. The effect will be a quickly-formed and evenly-tinned soldering iron.

To make an efficient joint, flux is first applied to and around the contact faces while the articles being joined are held firmly together. The heated bit and the solder are brought close to the joint on which a droplet of solder is allowed to form. The soldering iron is passed over and around the joint, using a light touch, and this transfers heat to the contact faces and also melts the solder, which flows evenly over and between the joint areas.

It is important that the following precautions are observed:

(1) The solder, soldering iron, and the parts being joined must be clean.
(2) Areas beneath and around the joint must be protected before soldering commences.
(3) The soldering iron must be of the correct shape and size for the job and be heated sufficiently to form a strong joint.
(4) The correct grades of solder and flux should be used.
(5) Solder must not be removed from the bit by flicking the iron or use of the fingers, excess solder should be removed carefully before using the iron again.
(6) The tinned surface of the bit must be touched with solder immediately after use to prevent removal by burning when the iron is re-heated.

19.5.1 Soldering irons

Soldering irons are produced in a variety of shapes and sizes. In modern electrical and telecommunications workshops the electrically heated instrument is most popular, especially for small work. For production purposes this type of soldering iron is often fitted with a magazine containing the flux-cored solder. These magazines look somewhat like fishing reels and the feed mechanism is operated by a trigger.

Diagrammatic sketches of some soldering irons are shown in Fig. 19.2.

19.6 Plumbing

For sealing joints and sleeves over 25 mm diameter, plumbing methods are adopted in place of soldering. This process entails shaping solder in its plastic state around the joint. This moulding of the joint is called *wiping* because two moleskin cloths coated with tallow and called the *catch cloth* and the *finishing cloth* are used.

The pot and ladle method of making joints is preferred to any other because it is quicker and naked flames are not used close to the joint.

The solder is heated in a pot which rests on a stand, the source of heat being blowtorches presented to apertures in the stand. The temperature of the solder in the pot is raised to about 400°C, a temperature which can be judged by dipping a piece of insulating paper into the solder. When the paper turns dark brown, without charring, the temperature is correct.

A dry, clean ladle is used to remove the scum from the surface of the molten solder

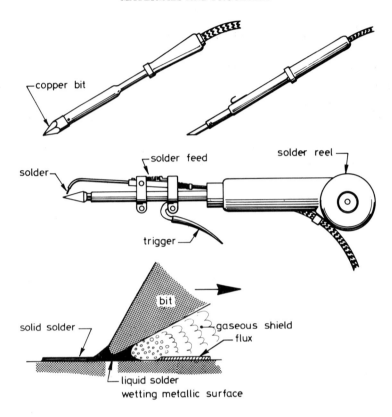

Fig. 19.2 Some soldering irons and the principle of soft soldering

and then the pot is taken to the cable. To limit the area covered by the wiped joint, *plumber's black* paste is applied to the areas around the cable sheath. This paste prevents solder from sticking to lead and reduces the amount of solder which would otherwise be used.

The metals to be joined are heated by pouring molten solder from a ladle over the sheath and sleeving, any excess solder being caught beneath the joint in the moleskin catch cloth. As the solder adheres to the joint areas and builds up, it is continually wiped to shape while still plastic using both cloths as necessary, any excess solder in the catch cloth being returned to the pot. The joint area is cleaned and allowed to cool, after which it is tested for air-tightness using air pressure from a foot pump and an adaptor inserted in a pressure test hole provided in the joint for this purpose.

The stick and blowtorch method of plumbing is used on small joints and where the pot and ladle methods are impossible. It consists of using sticks of solder and a blowtorch. The cleaned surfaces of the cable sheath and sleeve are heated by the blowtorch and the joint areas are tinned. When this has been done, successive amounts of solder are melted on the joint faces to form wipes, the blowtorch being used to keep the solder plastic.

19.7 Summary

Solders are low melting point alloys composed mainly of lead and tin. Solder must always have a melting-point lower than that of the metals being joined but higher than the expected service temperature of the joint. Eutectic solder melts and solidifies at a constant temperature of $183\,^{\circ}$C. Tinman's coarse solder has a small temperature range of solidification during which time it is plastic.

Fluxes can be chemically active killed spirits or protective resin, tallow, or oil. Their main purpose is to exclude oxygen. Soldered joints must be clean and kept free from oxide layers. Care must be exercised in the selection of the correct materials, heating to the right temperature, and safety procedures. Plumbing is carried out using pot and ladle, or stick and blowtorch, the soldered joints being built up and wiped to shape.

20. BRAZING AND WELDING

20.1 Brazing

Brazing or hard soldering describes the permanent joining of two metallic surfaces by introducing between them a thin film of brazing solder. In this respect soft soldering and brazing are similar, but the differences lie in the alloying elements of which the solders are composed, the temperature reached in forming a joint and the service temperatures of the articles.

20.1.1 Brazing alloys

Brazing alloys should conform to *BS 1845:1966*, which describes the whole range available to industry and includes the so-called silver solders.

Silver solders are alloys of silver, copper, and zinc and also in some instances cadmium. The fluxes used are dependent upon the melting temperature of the alloy. Below 760°C a fluoride-base flux is used, while above this temperature borax fluxes are preferred because they are more free-flowing when used at higher temperatures. Silver solders are used to join a number of different metals including carbon steel, copper, copper alloys and alloy steels.

Brazing brasses or *spelter* are alloys of copper and zinc and sometimes copper, zinc, and tin. The solidifying temperatures are high (over 850°C) and the joints are strong and durable. The flux used with spelter is generally of the borax type.

Self-fluxing brazing solders are available which are alloys of copper, silver, and phosphorus. They are always used in an oxidizing atmosphere which aids the formation of a very fluid and effective flux through the process of oxidation itself.

These alloys are not applied to ferrous metals owing to their tendency to form a brittle joint, and the temperature range lies between the silver solders and the brazing brasses. Table 20.1 gives the composition and solidification temperature range of a small number of brazing alloys.

TABLE 20.1
Typical brazing alloys

Typical composition (%)						Approximate solidification temperature range (°C)
Cu	Ag	Sn	Zn	Cd	P	
15	50	–	16	19	–	620 - 640
28	61	–	11	–	–	690 - 730
37	43	–	20	–	–	700 - 775
50	–	–	50	–	–	860 - 870
54	–	–	46	–	–	870 - 880
54	–	1	45	–	–	860 - 870
81	14	–	–	–	5	625 - 775
93	–	–	–	–	7	700 - 730

20.2 The brazing process

As in the case of soldering, brazing will be ineffective unless care is taken to prepare the articles to be joined. Cleanliness of the joint faces is essential and the removal of any traces of oil or grease most important. The contact areas should be filed where necessary and then rubbed with emery cloth. Carbon tetrachloride wiped across the joint faces will remove any grease.

The articles being joined may be clamped together using clips or iron-wire binding, although with fitting pieces this will not always be necessary.

The flux, which can be a thick paste of borax and water, is applied to the joint and the whole area is gradually heated to eliminate all traces of moisture. When this is achieved the heating can be more intense until the metal around the joint is white-hot. The brazing alloy is thrust into the flux and then brought into contact with the joint until it melts and flows. The heat supply, usually a blow lamp or blow pipe, is removed at this stage and the articles being joined should be enclosed by building around them a loose structure of firebricks, in order to conserve the heat.

Brazing can be carried out by dipping the fluxed joint areas in a bath of molten brazing solder, by placing the prepared articles in a furnace, and by using electrical resistance heating.

20.3 Welding

Welding is a process of uniting metals by fusing them together to form a permanent joint. There are two main types of welding process in common use, *fusion welding* and *pressure welding*. Fusion can occur by localized melting of the metals, or by pressure alone being applied to the heated joint faces without any melting taking place; an example of this is forge welding where two pieces of white-hot steel can be welded together by hammering.

In the fusion welding process extra metal is added to the joint area. This may take the form of a flux-coated consumable electrode or a hand-held filler wire. Pressure welding processes do not require additional metal to form a joint. The principles are shown in Fig. 20.1.

20.3.1 Pressure welding

Pressure welding can be divided into a number of different forms, excluding forge welding already mentioned.

Resistance welding is a combination of the use of pressure and the natural electrical resistance of the metals being joined. The articles are butted together and an electric current is allowed to flow either through the faces being joined, or near to them. The resistance to the passage of the current produces sufficient heat to make the metal plastic. The pressure is applied, the interfaces are pressed firmly together, and a sound weld is formed.

Spot welding is a method of resistance welding with pressure in which an electric current is passed, via electrodes, through the articles being joined which overlap one another. The resistance of the metals heats the joint areas until they become plastic, and the pressure between the electrodes forms the weld which is approximately the

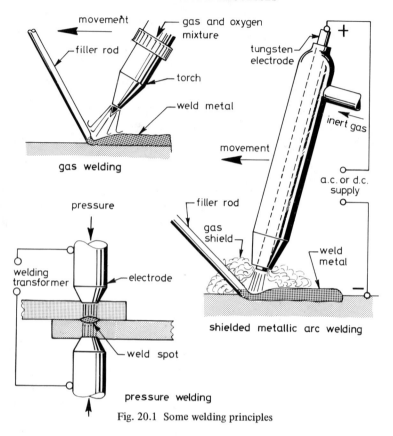

Fig. 20.1 Some welding principles

same size as the electrode tip. Spot welding is applied to relatively thin metals such as steel sheet. The welds are formed very quickly giving a high rate of production.

Seam and stitch welding are variations of the spot welding process in which wheel-shaped electrodes transmit the current and provide the pressure.

Stud welding is an electrical resistance welding technique in which the stud, which is a piece of screwed rod, is used as an electrode. An arc is struck between one end of the stud and the article to which it is to be attached. The arc produces the heat required to bring about localized plasticity in the metals and pressure is applied to the stud to complete the weld. This process is used to provide firmly anchored screwed rods in positions where bolts cannot be used.

Flash welding is the technique of joining together the ends of metal sheets and tubes. The ends to be joined are heated by striking an arc between them and when in a plastic state they are thrust together under pressure to produce the weld.

Fig. 20.2 shows diagrammatically some of the pressure welding processes and a small number of standard welded joints.

20.3.2 Fusion welding

Fusion welding describes a series of processes in which no pressure is used to

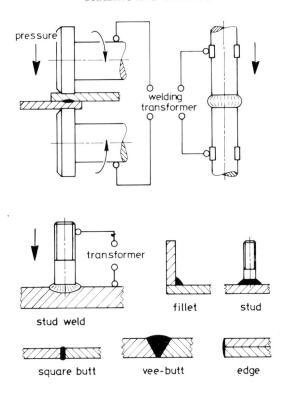

pressure

welding
transformer

transformer

stud weld

fillet

stud

square butt

vee-butt

edge

Fig. 20.2 Some pressure welding processes and standard welded joints

consolidate the weld. The metals being joined are brought to a state of fusion by heat generated either by a gas flame or an electric arc. Additional metal is deposited as a part of these processes. With *gas* and *carbon-arc welding* this is often in the form of a rod or wire, while with *metal-arc welding* the rod or wire becomes the electrode and as this melts it fills the space between the articles to be joined as the weld progresses. Welding rods often have a chemical composition similar to that of the parent metals.

Coated electrodes are frequently used in electric-arc welding. The coating may consist of a non-metallic sheath which melts ahead of the metal core and forms a protective flux around the joint and the weld metal. It is often chemically capable of reducing oxides and other impurities and inducing them into a slag which forms on top of the weld metal. Some electrode coatings produce, on heating, a protective inert gas which will exclude oxygen, and may also contain alloying elements to assist in strengthening the weld. This type of welding process is known as shielded-arc, because the electric arc operates within an inert gas shield.

Gas welding is a fusion process in which the surfaces being joined and a filler rod are melted by the flame from an oxy-acetylene torch. The mixture of oxygen and acetylene takes place in the torch, the amounts of each gas being regulated by valves attached to the gas cylinders. Temperatures up to $3000°C$ are possible with this process and the rapid melting ensures a quick weld without much loss of heat by conduction through the parent metal. An additional benefit is that distortion and oxidation of metal surrounding the weld are reduced.

Welding cast iron is confined to repair work and owing to the variety of cast irons available it is necessary to know the composition of the metal.

In general, cast iron can be gas welded or metallic-arc welded. In the former process it is preferable to use a filler rod made from Monel metal together with a powdered borax flux. In all cases of gas welding it is necessary to pre-heat the castings to avoid any rapid localized heating and cooling which could cause cracking of the parent metal. After welding, the castings should be allowed to cool slowly, either in a furnace or by covering them with sand or some other heat-conserving material.

Cast-iron repair welds can be made using iron, steel, or manganese-bronze welding rods. Steel or iron rods should have a high silicon content and pre-heating must be carried out. The main disadvantages with these materials is their high melting temperature and the extreme hardness of the weld metal which makes machining difficult. Manganese-bronze rods require less pre-heating time and the welding temperatures are reduced. A special flux is used in conjunction with these rods.

With metallic-arc welding of cast iron, Monel metal rods coated with a special flux are used. Pre-heating should be carried out where possible.

Welding of copper other than de-oxidized copper, presents some difficulties and should be avoided. De-oxidized copper can be welded using a filler rod containing pure copper and small amounts of silver and phosphorus. Pre-heating is essential because of the very high thermal conductivity of copper.

Brass welding presents its own special problem owing to the fact that the zinc content tends to evaporate and cause porosity. In addition, the zinc oxide vapour released during welding is poisonous to the operator and also clouds his vision of the work. Filler rods used with brass and other copper alloys sometimes contain aluminium as a de-oxidizer and silicon to increase the fluidity of the weld metal.

20.4 Summary

Brazing is a method of joining metals by introducing between them a thin film of brazing solder.

Brazing alloys should conform to *BS 1845:1966*.

Cleanliness is essential to the formation of a sound brazed joint.

There are two main types of welding: fusion welding and pressure welding.

In fusion welding extra metal is added to the joint area whilst in pressure welding no extra metal deposits are required.

21. BENCH WORK

21.1 The basic processes

Despite the amount of skill built into the large number of machines available, the basic individual workshop processes such as the use of the hacksaw, file, taps and dies, drills and reamers, hammer and cold chisel, scrapers, etc. will always have an important part to play in production engineering. This is especially true for the fitter, who makes more use of his knowledge and skill with hand tools than he does with machines.

Many of the processes described in this chapter are elementary, but it is these very basics that are so important to the production of pleasing and accurate work.

21.2 Measurement

It is always good practice to work to the dimensions on a sketch or drawing, because in this way shape and size can be pre-determined before the article is produced from metal.

A study of any drawing will show that it is made up of straight lines and curves only and the dimensions on the drawing relate to these and to angular displacement. Therefore, in order to translate a picture of an article into hardware, it is necessary to be able to make three different types of measurement on metal surfaces, linear measurement, non-linear measurement and angular measurement.

21.2.1 Linear measurement

Linear measurements are measurements along a straight line and relate to such things as the length and thickness of a piece of metal, the diameter of a rod or wire, the depth of a screw thread, etc.

The accuracy required in reproducing linear dimensions will determine the methods and instruments to be used. It is not possible to be absolutely accurate in reproducing a dimension but very accurate measurements are possible provided that the necessary equipment and instruments are available.

The engineer's rule. The engineer's rule, combined with a scriber and a surface plate, gives an accuracy of about 0·25 mm. This can be done by placing the rule and the article to be measured together on a surface plate as shown in Fig. 21.1. Resting both on a true plane surface removes the possibility of error at one face, leaving the judgement of the eye as the sole means of producing the required measurement at the other. The face common to both rule and article, from which the measurement is made, is called a *datum.*

An alternative method of producing the dimension is to use a scribing block on the surface plate as shown in Fig. 21.2. The scriber point is set to the dimension which is then transferred to the article. This still relies upon the accuracy of the eye in judging the measurement, but is slightly more accurate than the method previously described.

Calipers. Calipers are used in conjunction with the steel rule for measuring inside and outside diameters. Different designs of caliper are used for the two types of work

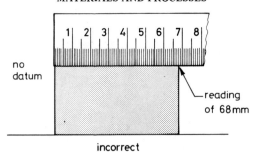

incorrect

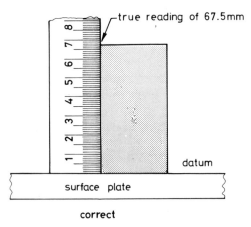

correct

Fig. 21.1 Measurement using a datum

but the principle of measurement is the same. The distance between the tips of the calipers is carefully adjusted until they are just touching the surface of the work. The calipers are then presented to a steel rule and the diameter is read off. Accuracy depends upon the feel when judging the correct setting across the tips of the calipers, more care and skill being required to gauge the feel of internal diameters than external diameters.

There are three possible sources of error in using calipers; judging the correct feel of the tips against the surface, making sure that a true diameter is being measured and reading the measurement across the caliper tips. Even so the degree of accuracy is the same as that using rule and scriber.

Calipers can be set to a checking dimension and then used to measure articles being machined without removing them from the machine tool. Fig. 21.3 shows sketches of both inside and outside calipers measuring diameters.

A *vernier caliper* consists essentially of a combined fixed jaw and rule and a sliding jaw. The sliding jaw carries the vernier scale while the rule is marked off in centimetres and decimal sub-divisions. A sketch of a typical vernier caliper is shown in Fig. 21.4.

Reading the vernier provides a means of removing the guesswork from the determination of a fractional part of a millimetre by the use of a mechanical device. This device is the vernier scale itself which is marked on the sliding jaw.

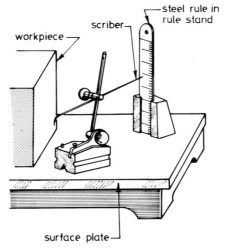

Fig. 21.2 Dimension-finding using a steel rule and scribing block

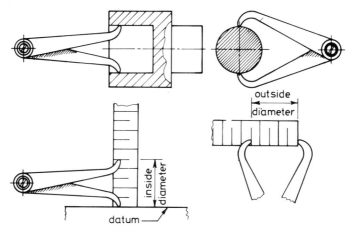

Fig. 21.3 Inside and outside calipers

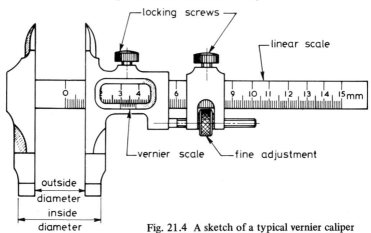

Fig. 21.4 A sketch of a typical vernier caliper

The fixed scale has main divisions of 1 cm sub-divided into tenths (1 mm) or twentieths (0·5 mm). The sliding vernier scale measures 9 mm over-all, marked into ten equal parts, each division being 0·9 mm. The accuracy obtainable is, therefore, the difference between the width of one sub-division on the fixed scale and the width of one sub-division on the vernier scale, that is 1 mm minus 0·9 mm, which is 0·1 mm.

The method of reading the vernier is as follows.

(1)   Read off the centimetres on the fixed scale.
(2)   Read off the millimetres on the fixed scale.
(3)   Read off the fraction of a millimetre indicated by the numbered vernier division that is coincident with a sub-division on the fixed scale.

Fig. 21.5 shows a selection of vernier readings.

In practice accuracy to 0·1 mm is insufficient for most engineering applications and so verniers are available which have an accuracy of 0·02 mm, to be on the safe side. On these the main scale is marked out in cm, mm, and 0·5 mm divisions, the vernier scale being 1·2 cm long divided into 25 equal parts. Most verniers have two scales, one on each side of the movable jaw. One reads internal diameters and the other external diameters.

An advantage of vernier calipers is that they can be made to extend over a wide range of dimensions in a single instrument.

The micrometer. This gives greater accuracy of reading than the vernier but has the disadvantage that a number of instruments are necessary to cover a wide range of dimensions.

The common micrometer will give an accuracy of 0·01 mm. Fig. 21.6 shows a sketch of a 0-25 mm micrometer, indicating the principal features and a series of readings. When measuring an article it is advisable to bring the spindle into gentle contact using the thimble and then turn the ratchet to its second click. In this way excessive clamping of the article between the spindle and the anvil will be avoided.

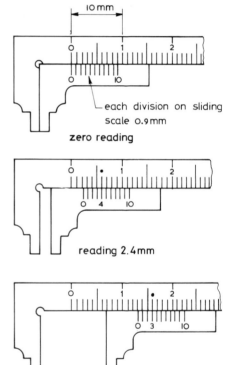

Fig. 21.5 Some vernier readings

Micrometers are available to measure external diameters, internal diameters, and depth. Micrometer devices are attached to some machine tools so that accurate positioning of the work or the cutting tools is possible.

When the datum line on the barrel does not coincide with the zero on the thimble,

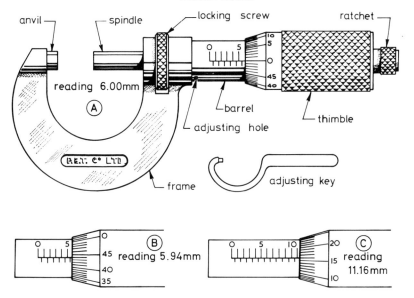

Fig. 21.6 A sketch of a micrometer and some readings

an adjustment must be made by revolving the barrel sleeve. This is done using the adjusting key.

When not in use the micrometer should be placed in its case and kept safely away from machine tables, lathe headstocks, etc.

The micrometer can be read using the following sequence.

(1) Read off the number of whole millimetres visible above the datum line on the barrel.
(2) Read off any half-millimetre division visible beneath the datum line on the barrel.
(3) Read off the number of divisions on the thimble, remembering that there are 50 in all.

Referring to the diagrams in Fig. 21.6, the reading at sketch A is made up as follows:

1. Above datum	5·00 mm
2. Below datum	0·50 mm
3. Thimble	0·50 mm
Total reading	6·00 mm

Sketch B shows a reading of 5·94 mm as follows:

1. Above datum	5·00 mm
2. Below datum	0·50 mm
3. Thimble	0·44 mm
Total reading	5·94 mm

Sketch C shows a reading of 11·16 mm as follows:

1. Above datum	11·00 mm
2. Below datum	0·00 mm
3. Thimble	0·16 mm
Total reading	11·16 mm

21.2.2 Non-linear measurement

Measurement of the non-linear features of an article presents special problems and requires skill, patience and first-class equipment. The measurements in this category include flatness, roundness, squareness, straightness and parallelism. Some of these are discussed in the following sections.

Flatness. This is determined in the workshop by comparing a surface with a known standard. A copper sulphate solution (engineer's blue) is smeared on a surface plate. The article being tested, which must be free from burrs, is pressed on this surface and moved across it in a figure-of-eight movement. The areas of contact of the article will be seen as blue spots. When the whole surface is blue the article is reasonably flat.

Roundness. The roundness of a cylindrical article can be checked by resting it in vee blocks beneath a dial indicator with which it is in contact. The dial indicator is zeroed and the article is slowly rotated. Any variation in roundness will immediately be registered by the pointer on the dial indicator. A sketch of this arrangement is shown in Fig. 21.7.

21.2.3 Angular measurement

A test for squareness between two faces is a measurement of an angle of 90°. This can be done at the bench by using the try-square, as indicated in Fig. 21.8.

To obtain angles other than right angles, to an accuracy of 0·25°, an engineer's protractor is used, and for very accurate measurement a vernier protractor is required. A sketch of an engineer's protractor is shown in Fig. 21.9.

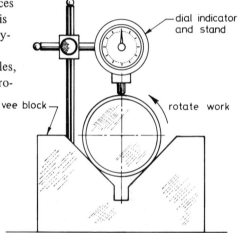

Fig. 21.7 Testing for roundness

21.3 Marking out

Marking out is the technique of producing scribed lines on the surfaces of articles so that they can be produced to the sizes and shapes laid down in the drawings. The equipment used in marking out is common to that described in the measurement section and the same amount of skill and care is necessary for both operations.

When a part has a number of faces it is advisable to make two of them at right angles to one another and ensure they are flat before proceeding with the marking out. Each face can then be used as a datum for all dimensions.

When an article is required that has no parallel lines or square surfaces, centre lines should be used as data. All the measurements would be referred to the centre lines during marking out so that accuracy of form relative to these lines is assured.

21.3.1 Equipment for marking out

The equipment for marking out includes surface plates, vee blocks, rules, try-squares,

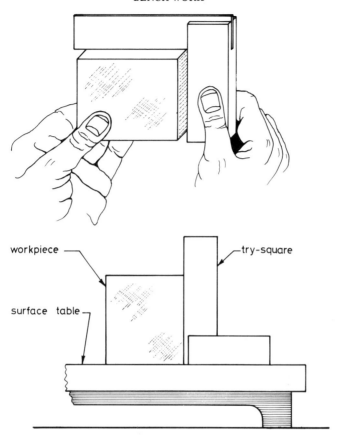

Fig. 21.8 Using the try-square

and the measuring devices previously mentioned. The scriber, centre punch, dividers, and odd-leg calipers are also required.

The scriber is made from high-carbon steel and is hardened and tempered at the points. It is capable of producing a clean visible line in bright mild steel without the need to use engineer's blue. The line should be produced by marking once only using a firm steady pressure.

There are two types of scriber, one which is held in the hand and used against a steel edge and the other which is mounted in a scribing block. Both of these are sketched in Fig. 21.10. When not in use the scriber points should be covered with corks to save them from damage and also for safety reasons. When in use only one of the corks should be removed.

Centre punches are used to assist in marking out the outline of an article, particularly on rough surfaces, and also for making start points for drills and engineers' dividers. The one used for drill centring has a 90° point while that used for divider points is more slender and has a 60° point.

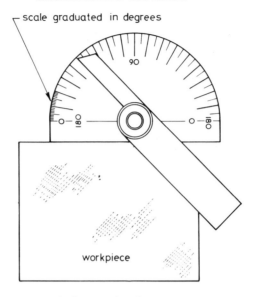

Fig. 21.9 An engineer's protractor

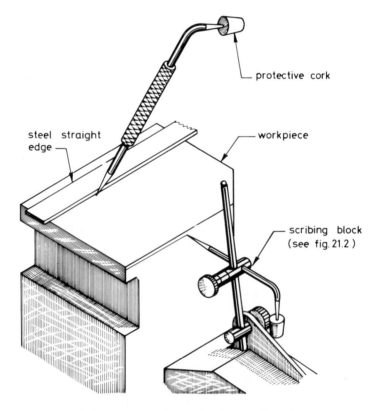

Fig. 21.10   Using two different types of scriber

Odd-leg calipers are instruments used to mark out parallel lines. The odd-leg rests against a machined edge along which it slides while the other pointed leg marks out the required lines. This is shown in Fig. 21.11.

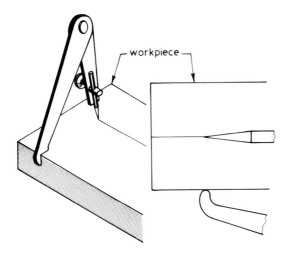

Fig. 21.11   Using odd-leg calipers

The scribing block is used in conjunction with the surface plate. The scriber is adjustable for height and angle, and a screw attachment is usually provided to give fine adjustment. Generally, the scriber should be kept horizontal with a minimum length of protrusion from its clamp. Measurements are taken from a steel rule for applications requiring minimum accuracy, but where this is important a vernier height gauge should be used.

In addition to the scribed lines, centre dots along the lines are often used to define an accurate outline. During machining, half the line and dots should be removed because the centre of the line represents the limit of the outline of an article.

21.4 The hacksaw

Hacksaws are used to cut metal to length or to cut out shapes in sheet metal and other relatively thin sections. The hacksaw consists of a frame, a handle and a removable blade. The frame may be adjustable, so that blades of various lengths can be fitted or fixed to take only one length of blade.

Metal removal should take place in the minimum time with minimum effort. This often requires patience and a great deal of care. When cutting out a shape it is advisable to saw to scribed lines and be within, say, 0·5 mm of these lines, leaving a minimum amount of metal to be removed by filing. Accuracy cannot be achieved with a hacksaw.

It is strongly recommended that all cuts should be made in a vertical direction so that downward pressure is always exerted in the forward cutting stroke. Thin sections must be rigidly held to prevent flexing of the metal being cut, and the full length of the saw blade should be used. Cuts should always be started downwards over an edge of the workpiece, never upwards against an edge. This is shown diagrammatically in Fig. 21.12.

Where all metal-cutting tools are concerned, priority should be given to personal safety. Cutting edges must not be touched with bare hands and tools must never be placed so as to create a safety hazard.

Saw blades have teeth that are very hard, and cannot be sharpened or re-set when they wear down. The setting of the teeth on a new blade is achieved either by having alternate teeth bent slightly outwards to either side, or by forming a wavy pattern on the cutting edge before the teeth are machined. This makes the cut slightly wider than the blade and prevents the blade from sticking or becoming tight in the cut.

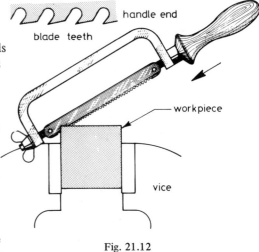

Fig. 21.12
Correct use of the hacksaw when starting a cut

Hacksaw blades can be obtained hardened and tempered all over, with the teeth only hardened and the back of the blade soft or, in the case of blades having teeth on both edges, with the centre of the blade soft.

21.5 Filing

Filing is an important and difficult art which, once mastered, can be used to produce work to an accuracy of 0·25 mm. The most fundamental points to observe are detailed below, assuming the operator to be right-handed.

Correct stance is most important for comfort and efficiency. The feet should be spaced well apart, with the left foot in front of the right. The body should feel comfortable and well balanced.

The working height can be judged by observing that the vice should be at the same level as the operator's elbow when he takes up his filing stance.

The handle should be held in the right hand in the same way for all filing operations. The position of the left hand varies according to the class of work and the type and size of the file.

To remove large quantities of metal in the shortest possible time the tip of the file should be held firmly in the left hand, palm resting on top of the blade, thumb across the top and fingers curled underneath. This grip enables the maximum force to be applied to the work.

For more accurate work and when using smaller files, different left hand grips are adopted. For curved surfaces the tip of the file should be held between thumb and index finger, curling the other fingers underneath the file so that the tips only are in contact with the surface of the file.

Where the fingers cannot be accommodated beneath the file they can be pressed on the top surface, with the thumb spread as far as possible from the fingers in such a way that there is an even distribution of pressure over all parts of the file.

The forward stroke only is the cutting stroke and requires the maximum force; the return stroke should be light with the file remaining in contact with the work.

The recommended methods of holding a file are shown diagrammatically in Fig. 21.13.

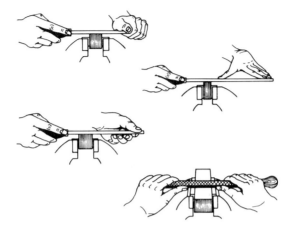

Fig. 21.13   Methods of holding the file

File positions relative to the work are most important. For thin sections the file should lie at a slight angle to the work surface so as to cover the maximum area during a cutting stroke.

For thicker sections the file should be held at a slightly larger angle to the work, the file being pushed across and along the surface at the same time. The direction of filing should be changed occasionally so that the filing marks are evened out over the whole surface. Fig. 21.14 shows the approximate file angles and the direction of movement during filing.

To obtain accurate work on thicker sections, tests should be carried out at intervals to ensure squareness and flatness. A try-square can be used to gauge the squareness between an edge and a surface and, by using the top edge of the blade, to give an indication of straightness. Holding the workpiece and try-square together against a light source will test these two features. When no light is seen between try-square blade and work surface, the required conditions have been achieved.

When greater accuracy is required, the surface plate and engineer's blue should be used to determine any high spots which can then be filed down.

21.5.1 Files

Files are made from high-carbon steel and are hardened and tempered along the cutting length of the blade. The *tang* or tapered end that is forced into the wooden handle is tempered to be tough and not hard so that it will not snap off when in use.

There is a wide range of files available varying in section and length, and each one has its own special use. The blade is cut across straight or at angles so that a large number of cutting edges are formed.

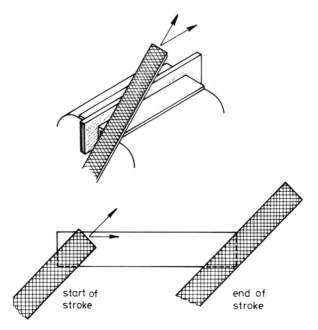

Fig. 21.14   File angles and movement relative to the work

New files should be used first on the softer metals and then passed for use on the harder metals. This is because the softer metals are more easily and efficiently removed using a new cutting surface and the teeth will retain sufficient sharpness to be used on the harder metals.

The teeth of files can become blocked with metal particles and so a wire brush or file card should always be available to remove these. If this is not done the surfaces of the workpiece will be damaged and the efficiency of the file reduced. A piece of chalk rubbed across the teeth of the file will help to prevent metal pick-up; paraffin can be used for the same purpose when filing aluminium.

21.6 Hand scrapers

Scrapers can be used after filing or machining to produce working surfaces that are flat or round. For flat work there are two types of scraper, one operated by pushing, the other by pulling. These are shown in Fig. 21.15.

Hand scraping is a time-consuming and costly process requiring great skill and patience. It is also a necessary method of obtaining flat surfaces that slide one upon the other and is widely used in the machine tool industry and for large diameter bearing sleeves. In addition to the provision of accurate surfaces, the minute indentations left by the scrapers form tiny pockets for the retention of oil, thus increasing the life of the sliding parts.

A scraper is made from high-carbon steel; the cutting edges are very hard and must be kept sharp. The tools remove a tiny sliver of metal with each cutting stroke.

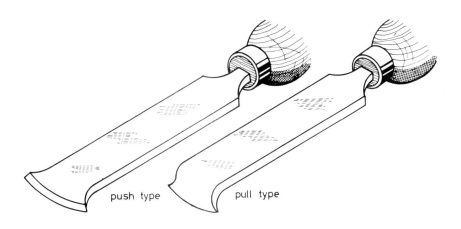

Fig. 21.15   Basic types of scraper

21.7 Cold chisels

Cold chisels are made from octagonal-section high-carbon steel bar and are hardened and tempered at the cutting end. There are four principal cold chisels in common use, these are the flat, the cross cut, the round nose and the diamond point.

The *flat chisel* is used for cutting relatively thick sheet metal along a line or a row of holes, removing rivets and rusted-in nuts and bolts, and removing metal after other operations have been completed. One disadvantage of using a flat chisel for cutting sheet metal is that distortion is likely unless great care is taken in supporting the work.

The *cross cut chisel* is made so that its cutting edge is wider than the rest of the blade. Its principal uses are in cutting narrow slots up to 12 mm wide and grooving metal which is to be chipped away over a large area using a flat chisel. The *round-nose chisel* is used to make oil ways in bearing sleeves and shafts. The *diamond-point chisel* is used to cut vee grooves and to ensure sharp internal corners in slots and keyways. Sketches of the four types of chisel are shown in Fig. 21.16.

21.8 Summary

Bench work requires a great deal of skill, knowledge and patience. The measurements that must be made include linear measurement, non-linear measurement, and angular measurement.

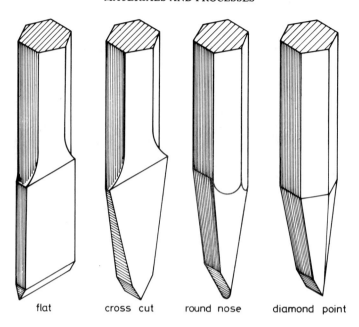

flat   cross cut   round nose   diamond point

Fig. 21.16   Four types of common cold chisel

Calipers are used to measure internal and external diameters. A vernier caliper is a device for taking linear measurements to a degree of accuracy of 0·02 mm. A micrometer gives greater accuracy than a vernier, reading to within 0·01 mm. Marking out is the process of producing scribed lines on surfaces that conform in shape and size to outlines on drawings.

22. THE DRILLING MACHINE

For most engineering applications a drilling machine is essential to generate cylindrical surfaces in materials being worked. Sometimes portable drilling machines powered by compressed air or electricity are available. Parts to be drilled are often made from comparatively thin metal and the pressure required to penetrate the work is provided by the operator. When smaller, thicker sections are involved the work is taken to the drilling machine. Where accuracy is essential the hole centres should be marked out and centre-punched to assist in the location of the tool. For extreme accuracy, a rigidly-held hardened steel bush or sleeve is used to guide the cutting tools.

The *sensitive drilling machine*, sketched in Fig. 22.1, shows some of the essential features of this robust machine tool. It is called sensitive because the downward pressure or feed of the cutting tool is provided by the hand of the operator via the capstan wheel. A device called a *chuck* is fitted to hold the cutting tool, the centre line of which must from all directions be at 90° to the top surface of the adjustable worktable. An electric motor is used to provide the power to rotate the tool, and a range of speeds is available from the cone pulleys, enabling the best speed for the size of tool and class of work to be selected.

Some means of clamping articles to the worktable should be provided, but this is not always necessary because the work can be clamped using a machine vice. Sometimes the mass of the workpiece itself is sufficient to ensure steadiness during the cutting process.

22.1 Cutting tools

The main purpose of any drilling machine is to rotate a cutting tool to generate holes for rivets, bolts, pins, etc., but it is also designed so that other tools can be fitted for special purposes.

The *twist drill* is made from circular-section tool steel and the main body is formed by cutting into it helical *flutes*. The leading edge of each flute is sharpened into a cutting edge and this is backed up by a narrow, flat strip called the *land*. The flutes are designed to give the required cutting angle and allow for swarf removal.

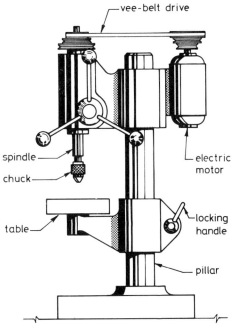

Fig. 22.1 The sensitive drilling machine

The body of the drill is hardened and tempered while the *shank*, or plain portion which fits into the drilling machine, is relatively soft but tough. Shanks are usually parallel for small sizes and tapered for the larger sizes. The tapered shank ends with a flat *tång* which locates with a slot in the drilling machine spindle. This provides the positive drive necessary to penetrate thick sections of metal.

The point of the drill is formed by grinding the end of the flutes to an included angle of 118°, with a 5° clearange angle sloping back from the cutting edge. The angle of twist cannot be altered by re-grinding but the cutting and clearance angles can be destroyed by faulty grinding. Oversize holes are produced when the point of the drill is offset from the centre line, and when the clearance angle is less than 5°, the drill will not feed into the work.

Rotating the drill at too high a speed causes excessive wear and overheating which can result in crumbling of the cutting edges. The correct speed at which the drill should rotate is important and depends upon its diameter and the material being drilled. In most cases it is necessary to apply a combined coolant and lubricator of some kind to ease the passage of the drill through the work and prevent overheating. Table 22.1 gives the approximate cutting speeds for carbon steel twist drills cutting three different types of metal with an ample supply of lubricant.

TABLE 22.1
Approximate speeds for carbon steel twist drills

Diameter of drill (mm)	1·5	3	5	6	8	10	12	15	20	25	Approx. cutting speed (m/min)
Rev./min for mild steel	1830	900	600	450	350	300	200	175	140	120	9
Rev./min for brass	9000	4500	3000	2250	1825	1500	1000	900	675	575	45
Rev./min for annealed cast iron	4250	2100	1400	1000	850	700	500	400	325	250	21

A *reamer* is a cutting tool rather like a drill in appearance. It is used to finish drilled holes to size and also to make parallel holes into tapered holes. Reamers are made from tool steel and are hardened and tempered in the same way as twist drills. There are hand reamers for use at the work bench and machine reamers which fit into drilling machines.

Hand reamers have parallel shanks and are provided with a square end to fit a hand wrench, while machine reamers have straight flutes that may be parallel or tapered and others have helical flutes, the helix angle of which is smaller than that found on a twist drill.

Fig. 22.2 shows a small number of drills and reamers.

Countersinking is a method of enlarging holes at their ends. For general countersinking purposes a drill twice the diameter of the hole is used, but when soft metals are being worked, special countersinking drills are often employed to make depressions for screw heads and rivets. These are shaped to the required angle, as shown in Fig. 22.3.

To accommodate cheese-headed screws beneath the work surface, a flat countersinking drill is used. This has a pilot pin in the centre of the flat cutting edge which

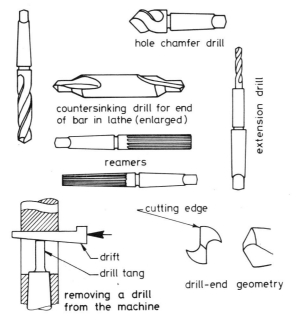

Fig. 22.2 Drills and reamers

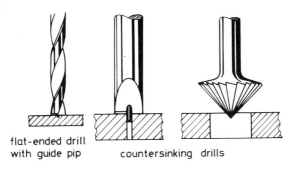

Fig. 22.3 Drills used for special purposes

locates in a previously drilled hole or a hole containing a screw thread. An example of this type of drill is also shown.

22.2 Summary

Drilling machines generate cylindrical surfaces in workpieces. The machine must be robust in construction, rigid, and safe to use, and the tools correctly ground and sharp.

Twist drills are relatively weak tools which will bend or break when excessive downward forces are applied. Reamers are designed to produce accurate surfaces in previously drilled holes. Tools used in the drilling machine have a best speed for each type of material and each size of tool.

The greatest attention must be paid to the safety aspects of the use of the drilling machine.

23. TAPS AND DIES

23.1 Screw cutting

Taps and *dies* are items of metal-cutting equipment used to produce screw threads. Internal threads are cut using taps, while external threads are cut using *stocks* and *dies*. Although screw threads are also produced on lathes and other machines, the processes and equipment in this chapter will be confined to those used in benchwork.

23.1.1 Tapping

Tapping a threaded hole requires a great deal of care and patience. First it is necessary to drill a good quality hole using a tapping size drill. The diameter of this drill is approximately equal to the smallest diameter of the thread.

The next step is to produce a small countersink at the open end, or ends, of the hole. This will prevent burrs forming during the tapping process.

It is usual to employ three different taps to produce one tapped hole. The first one, called the *taper tap*, makes the roughing cut. It is heavily tapered at the end which enters the hole and increases gradually in diameter, reaching the maximum size only over the last few teeth. The next tap is called the *second tap* and is less tapered at its free end. The final operation is carried out using the *plug tap* and this achieves the required thread size and a good finish.

It is essential to ensure the taps enter the work at 90° to any part of the top surface of the metal. This can often be achieved using a try-square as a means of checking. When tapping thin metal the problem of keeping the tap vertical to the workface becomes even more difficult. A means of overcoming this problem is to clamp a nut, of the same thread size as the one to be cut, over the hole, so that the taps pass through the nut before entering the work.

The tap wrench must not be rotated continuously, as this will cause the tap to jam in the hole, trapping the metal chips and damaging both tap and hole. Initially, the tap should be rotated forward a quarter of a turn, then back a quarter of a turn, forward half a turn, back half a turn, continuing in this way for the whole depth of the thread. Winding the tap out of the hole after each bite of the cutting edges removes the metal chips. During the tapping operation lubricant must be liberally supplied to the cutting tools.

Taps are made from high quality cast steel and are hardened and tempered all over except for the square end over which the tap-wrench fits. The extreme hardness of the cutting teeth makes the taps brittle, so great care must be taken to get an even pressure as tapping progresses.

The cutting edges are formed by straight flutes which also allow for metal removal. Fig. 23.1 shows a set of taps and a tap wrench.

23.1.2 Dies

Dies generate external threads and it is essential to ensure that the outside diameter of the work is correct for the thread being cut. A slight taper should be provided at the starting end of the work to allow the dies to gradually cut away the metal.

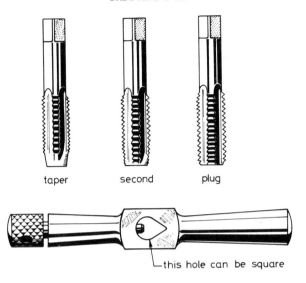

taper   second   plug

this hole can be square

Fig. 23.1 Sketches of a set of taps and a tap wrench

Dies are of two main types: a die nut, cut in two halves and designed to fit into the stock, and a split die, made in the shape of a ring split at one point by a vee groove. The latter is drawn in Fig. 23.2.

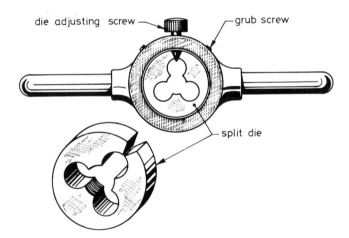

die adjusting screw   grub screw

split die

Fig. 23.2 A die-stock and split die

The split die is opened out by the wedging action of a set screw, the end of which engages in the vee groove. This type of die is used to cut the smaller sizes of screw threads, the larger sizes being cut by the halved die set. The relieving holes drilled in the die form the cutting edges and allow for the removal of metal chips.

As in the case of tapping a thread, a roughing cut is taken first and the size of the die nut is gradually reduced until the correct thread form is achieved. When the die has run the required thread length on the bar at any one setting it must not be tightened on the return stroke, because the cutting action takes place only in one direction. Lubrication should be liberally supplied to the workface to reduce friction and prevent overheating and loss of temper in the dies.

23.2 Summary

Taps and dies are used to cut internal and external threads respectively. Thread-cutting operations require the liberal use of lubricants. Taps must always be at 90° in all planes to the surface of the work.

24. THE LATHE

24.1 General description

Lathes are machine tools in which material is rotated; while cutting tools generate cylindrical shapes, tapers, and helices. The methods of holding and supplying the work to the machine, and the way the cutting tools are fitted and presented to the work can vary but the basic principle of *turning* remains the same for all lathes.

Lathes are available in various sizes depending upon the class of work they are designed to produce. All must be solidly built, rigid structures and the finish of all slideways, fixed, and moving parts should be of the highest accuracy. Devices are provided to hold the work firmly against the forces set up during machining. The tools themselves must be clamped in an arrangement which can traverse the lathe bed parallel to the centre line of the work, provide for movement across the bed at 90° to the centre line (tool feed), and also a combination of these movements to produce tapers.

Sufficient power must be available to rotate the work during cutting and drive the tool-traversing mechanism, both of which require variable speed devices so that the most suitable spindle speed for any given diameter of workpiece can be selected by the operator.

For screw cutting, the movement of the tool parallel to the work relative to the rotational speed must be capable of accurate determination.

24.2 Particular features

Fig. 24.1 shows a sketch of a typical lathe and indicates some of its principal features. The diagram also shows two methods of mounting work for turning.

The *height of centres* is the vertical height from the lathe bed to the centre line of the lathe. This distance determines the maximum size of work which the lathe will hold. The swing of a lathe relates to the maximum work diameter which can be accommodated. The *distance between centres* represents the maximum length of bar that the lathe will accept.

24.2.1 The headstock

The headstock is the powered end of the lathe containing the main drive to the lathe spindle, coned pulleys to vary the spindle speed, and the train of gears that drive the leadscrew. A method is provided of linking the gear wheel on the spindle shaft with the gear wheel attached to the leadscrew, so that the rotation of the spindle can be directly linked to the linear movement of the tool. A gearbox is fitted for this purpose which also provides a selection of speeds for various sizes of threads to be cut.

24.2.2 The tailstock

The tailstock is at the opposite end of the lathe bed from the headstock and can be moved in guideways along the bed to accommodate different lengths of work. The main purpose of the tailstock is to support work being turned between centres. To do this, the work being turned must have countersunk holes drilled in each end to locate

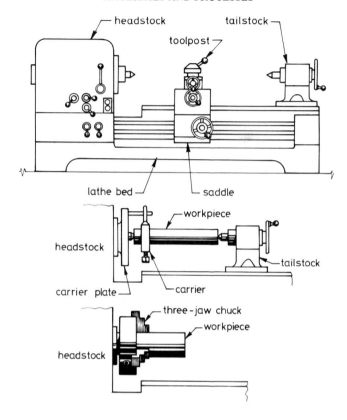

Fig. 24.1 A centre lathe and methods of holding work

on the pointed centres protruding from the tailstock and headstock. Grease is applied to the parts at the tailstock end because the centre remains stationary or 'dead' while the workpiece revolves. Once in position the tailstock is clamped rigidly to the lathe bed. The dead centre can be wound backwards or forwards by rotating the handwheel at the rear of the tailstock.

The tailstock is also used to drill axial holes in workpieces mounted on the headstock end of the lathe by replacing the dead centre with a drill and feeding it into the work, using the handwheel.

24.2.3 The saddle

The saddle, which fits over the lathe bed, is provided with guideways which give location and also enable it to move backwards and forwards parallel to the centre line. Mounted on the saddle are the means of sliding the cutting tools in different directions relative to the work. These are the *cross slide* and the *compound slide* situated one above the other, as shown in Fig. 24.2. On top of the slides is the *toolpost* itself, where the cutting tools are set in the correct position and clamped prior to any turning operation. The part of the saddle covering the front of the lathe is called the *apron*. This contains the traversing and screw-cutting mechanisms.

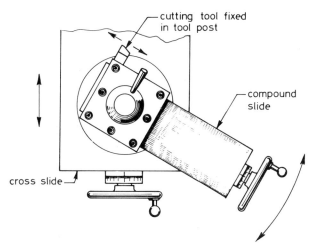

Fig. 24.2 The cross slide and compound slide

24.3 Work holding

Very long workpieces, or those requiring additional work and inspection after turning, are supported between centres as previously indicated in Fig. 24.1. In some cases it is necessary to provide the additional support of a *lathe steady*.

For workpieces of comparatively short length, which can be machined in one setting, the self-centring *three-jaw chuck* is used.

Turning between centres requires countersunk holes at each end of the workpiece. The holes must be accurate, have a smooth finish, and run true on the centres. Once mounted in the lathe the work must be free to rotate and not clamped rigidly by the tailstock. To drive the workpiece a *carrier plate* is attached to the driving spindle, and the *carrier* is tightened on the workpiece by means of a screw which must not damage the surface of the work.

One advantage of turning between centres is that the work can be removed and replaced, yet still run true to the original setting. This enables work to be reversed for further turning or passed to other machining or inspection processes having centres fitted as a means of location.

24.3.1 Lathe steadies

Lathe steadies are used as additional support for round bars of extreme length. When support is not provided, the forces exerted at the cutting tool will bend the workpiece and the turned diameter will vary along its length.

There are two principal types of lathe steady, one fixed and one attached to the lathe saddle which travels with the tool. A typical example is shown in Fig. 24.3.

The base of the steady is made to fit the top surface of the lathe bed and a clamping device is fitted. The whole structure must be firm and rigid.

24.3.2 The three-jaw chuck

The three-jaw chuck is self-centring and uses the three-point principle of location to

hold and locate the workpiece. The
jaws can be moved inwards or outwards
together by the operation of a square-
ended key, the amount of movement
being the same for each jaw. The inside
surface of the jaws is serrated to assist
in gripping the workpiece, while the
outside is stepped so that the jaws can
be reversed in action and used to hold
pipes and tubes from the inside. This
is shown diagrammatically in Fig. 24.4.

One disadvantage of the three-jaw
chuck is that work having a number of
concentric diameters can only be
machined in a single setting. It is
almost impossible to re-chuck a job
and set it to run true to the original
setting.

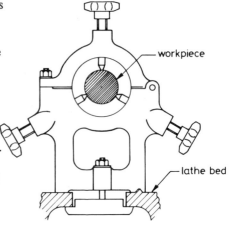

Fig. 24.3 A typical lathe steady

24.4 Turning

The action of the turning tool is one
of *shearing* the material, an action
which must be as smooth as possible to
obtain the best results. The position and
setting of the tool relative to the work
is very important.

The cutting point or edge of the
tool must be on the centre line of the
work and there should be the minimum
amount of overhang between the tip of
the tool and its restraint. Excessive
overhang leads to tool breakage or
vibration which gives a very poor
finish. Positioning of cutting tools is
shown in Fig.24.5 and further information on their operation is given in Chapter 25.

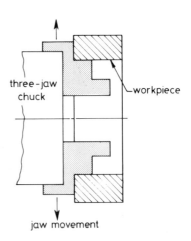

Fig. 24.4 Gripping the inside of a tube

Metal removal should be carried out on the principle of removing the maximum
amount of material in the shortest possible time. This does not mean that huge cuts
can be made at tremendous rotational speeds, because heavy cuts involve a concen-
tration of forces at the tool which will not produce good quality work.

There are, generally, two stages in the process of metal removal. The first one is the
roughing cut, which removes a large amount of metal very quickly, by choosing the
maximum speed and feed for the material being cut. The second is the *finishing* cut,
which removes a smaller amount of metal (say 0·5 mm). The finishing operation
requires less cutting force, the workpiece must be brought to size, and the finish must
be good. This is achieved by changing the tool and increasing the speed and reducing

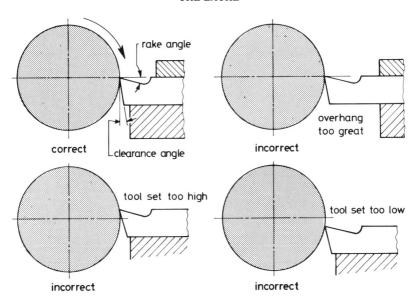

Fig. 24.5 Correct and incorrect settings of a lathe cutting tool (see 25.1.4)

the feed. Finally, all sharp edges must be removed from turned components, for safety and assembly reasons.

Throughout all cutting operations some form of coolant and/or lubricant should be supplied to the cutting tool to ensure a long life and reduce the frictional forces.

Safety is of the utmost importance and special attention must be paid to this when operating a lathe. Apart from wearing protective hats, etc., spanners, micrometers, and other measuring devices must never be left lying on the lathe during turning operations. Swarf removal must be carried out with care, both during and after using the machine.

24.5 Summary

Lathes are machine tools which rotate workpieces, while cutting tools generate cylindrical shapes and screw threads.

25. CUTTING TOOLS AND LUBRICANTS

25.1 Tool geometry

All cutting tools must have a basic wedge shape. The apex of the wedge forms the cutting edge, while the thicker part provides the strength and rigidity.

The action of all cutting tools is a shearing action and, to keep the length of the shearing plane as short as possible and reduce to a minimum the effort required, a *rake angle* is necessary. The rake angle on a tool depends upon the type of metal being cut and the material from which the tool itself is made.

Fig. 25.1 shows chisels cutting mild steel. The first drawing depicts a chisel with no rake angle; the second drawing indicates the correct angle for mild steel, and the third

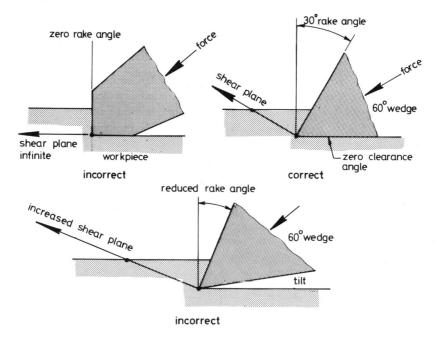

Fig. 25.1 Chisel edges cutting mild steel

drawing shows the effect of tilting a correctly ground chisel. In the first example, the length of the shear plane is the length of the workpiece and no shearing action is possible. In the second example, the chisel is set correctly to the work, the shear plane length is short, and the removed metal can flow easily over the face of the tool. In the third example, tilting the chisel has resulted in a reduction of the rake angle, the shear plane length is increased, and the chisel will dig into the surface of the workpiece. This means that, in addition to providing a rake angle, the tool must be set correctly relative to the work for effective metal removal to take place.

A clearance angle is essential with most cutting tools. This angle is formed on the tool to eliminate rubbing action between the tool and the workpiece. When rubbing does occur frictional forces are set up which consume power. Clearance angles are not provided on cold chisels because the cutting action relies on one face of the tool being in contact with the workpiece. No frictional problems arise as the cutting action is intermittent and there is no chance of the face-to-face contact causing overheating.

25.1.1 Hacksaw and file teeth

The hacksaw and file-teeth are multi-point tools having a number of teeth in engagement with the workpiece at any one time. A close inspection of a single tooth on a hacksaw and file blade will show that they are similar in shape and are given both rake angle and clearance angle. The teeth conform to the basic wedge shape and the radiused grooves ensure strength and space for removed metal.

The clearance angle is provided to reduce the friction between tool and work. This also reduces the power required during the cutting stroke.

Fig. 25.2 shows enlarged details of typical file and hacksaw teeth.

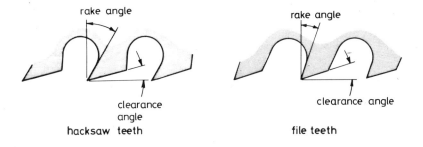

Fig. 25.2 Enlarged diagrams of hacksaw and file teeth

25.1.2 Twist drill

The twist drill is a single-point cutting tool in which the rake angle is equal to the helix angle of the flutes. The flutes also provide a means of carrying away the removed metal which spirals upwards out of the hole.

Fig. 25.3 shows details of a twist drill, indicating the cutting edges backed up by the land which provides strength and also allows the cutting edges to be sharpened. Sharpening of cutting edges and drill point must only be carried out using equipment specially designed for the purpose. Incorrect grinding of a drill will produce malformed and badly finished holes.

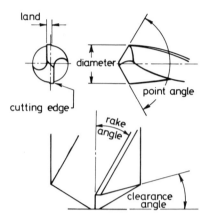

Fig. 25.3 Twist drill details

25.1.3 Taps and dies

Taps and dies have cutting edges with rake angles formed by the straight flutes on the tap and the holes in the die being machined off-centres (see Fig. 25.4). No clearance angle is possible so there is a frictional resistance to overcome during thread cutting. The exercise of great care and the liberal use of lubricant will reduce considerably the effort required to produce a thread and the possibility of breakage or damage occuring.

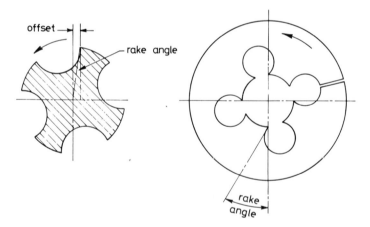

Fig. 25.4 Rake angles on a tap and die

25.1.4 Lathe tools

Lathe tools have rake and clearance angles as indicated in Fig. 24.5, which also shows the correct and incorrect setting of a tool relative to the centre line of the work. No purpose is served in carefully grinding the correct angles on a tool and then setting it incorrectly and destroying their effectiveness. Setting the tool *above* the centre of the workpiece *increases* the rake angle and *decreases* the clearance angle. This increases the possibility of rubbing, requires an increase in power, and will cause damage to the tool and the workpiece.

Setting the tool *below* the centre of the workpiece *decreases* the rake angle and *increases* the clearance angle. This setting will remove the strength of the tool and break away the cutting edge. The metal removed will not flow over the tool and the workpiece will have a poor finish.

Fig. 25.5 shows diagrams of a small number of lathe cutting tools and their positions when cutting. The direction of feed and tool travel are indicated together with the rake angles. Table 25.1 shows a small selection of rake angles for particular cutting tool and workpiece materials.

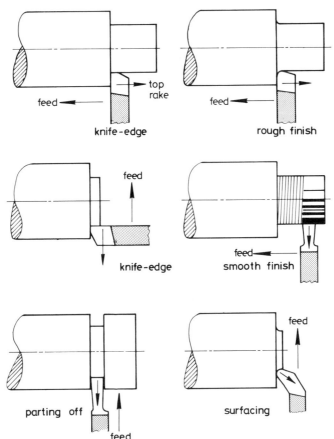

Fig. 25.5 Sketches of some lathe tools showing direction of feed and tool top-rake

Sharpening lathe tools becomes necessary after a time owing to normal wear, accidental damage, etc. For effective cutting the tool must be kept as sharp as possible and so it is sharpened on an *off-hand grinder*. This machine consists of a pedestal at the top of which is a motor driving two grinding wheels which protrude from either side of the machine. Small adjustable platforms are provided on which to rest the tool and present it at the correct angle for sharpening. Transparent plastics guards are fitted

TABLE 25.1
Rake angles for some cutting tools

Workpiece material	Tool material		
	HSS	SHSS	Stellite
Mild steel	20°	20°	20°
Grey cast iron	up to 10°	up to 80°	11°
Brass	0	0	8 - 12°
Copper	35°	35°	10 - 20°
Aluminium	40°	40°	10 - 20°

HSS = High-speed steel
SHSS = Super high-speed steel

to cover the grinding wheels and so protect the operator from flying particles (when guards are not available, protective goggles must be worn).

The grinding wheels have different textures, one is for roughing and the other for finishing. Assuming the wheels are running true, which is very important, the tool is presented first to the roughing wheel and then to the finishing wheel. Because the sharpening causes considerable heat to be generated, it is advisable to plunge the tool frequently into coolant. It is bad practice to allow a tool to become very hot and then quench it. This would cause minute cracks to appear in the surface of the metal which might lead to tool failure during any subsequent turning operation.

It is often sufficient to re-sharpen a tool by presenting it to the finishing wheel only. The angle of the face being ground must be maintained at all times. To avoid local wearing of the grinding wheels the tools should be moved backwards and forwards across their faces. This will preserve the true running of the wheels and prolong their useful lives. Fig. 25.6 shows diagrams of a tool grinder and a tool presented at its correct angle to a wheel for re-sharpening.

25.2 Cutting tool materials

Cutting tool materials must be very tough, harder than the metals they have to cut, able to operate at high temperatures without failure and capable of removing metal at a rapid rate.

25.2.1 High-carbon steel

High-carbon steel is made into simple lathe tools, cold chisels, files, scrapers, taps, and some hacksaw blades. The steel contains between 0·8 per cent and 1·4 per cent carbon. It is relatively cheap, possesses toughness, and can be hardened, retaining its useful properties at temperatures up to 250°C.

Chisels are made from steel containing between 0·8 per cent and 0·9 per cent carbon; thread taps and files from steel containing between 0·9 per cent and 1·1 per cent carbon; and scrapers from steel containing up to 1·4 per cent carbon.

25.2.2 High-speed steel

High-speed steel, as the name implies, was developed to allow higher cutting speeds and feeds to be used. This material is an alloy steel containing, in a typical composition; 18 per cent tungsten, 4 per cent chromium and 1 per cent vanadium. It is a fairly

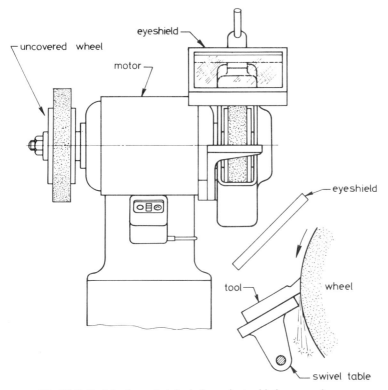

Fig. 25.6 Sketch of a pedestal grinder and a tool being ground

expensive material owing to its constituents and method of production. It possesses an acceptable standard of toughness, hardness and wear resistance, maintaining these desirable properties at temperatures up to 600°C.

High-speed steel is used to make twist drills, lathe turning tools and milling cutters, and can cut all but the hardest steels.

25.2.3 Super high-speed steel

Super high-speed steel has a composition similar to ordinary high-speed steel but contains, in addition, up to 12 per cent cobalt.

This range of materials hardens well, possesses an acceptable standard of toughness, and has good resistance to wear. The tools made from this alloy are used to machine the hard, high-alloy steels and for rapid production of repetitive parts.

Because super high-speed steel is costly to produce, small tips are made that can be electrical-resistance-welded to a mild steel shank. In this way the extremely hard tool bit is supported by a tough and softer metal, reducing not only the cost of production but also the possibility of breakage.

25.2.4 Other materials

Other materials, mainly in the form of tips, are as follows: *Stellite*, an alloy of chromium, cobalt, and tungsten; cemented carbides, such as *tungsten carbide* and *titanium carbide; ceramic* materials and *diamonds.*

25.3 Coolants and lubricants

All machining processes involve the shearing of layers of metal from the surfaces of workpieces, and during these operations large quantities of heat are generated. In addition, smaller quantities of heat are generated as the removed metal chips flow over the face of the cutting tool and from the friction between the flanks of the tool and the workpiece.

The harmful effects of this heat must be kept in check or the cutting edges of tools will rapidly wear away, soften appreciably and become useless, or become welded to metal chips. There is also the danger of expansion and distortion of the work surface with a consequent loss of dimensional accuracy and a poor quality finish.

A liberal flow of liquid substances, called *cutting fluids*, directed at the cutting face reduces these dangers to manageable proportions by providing lubrication and a ready means of carrying away the heat. In this way the tool can maintain its essential properties of a keen edge and extreme hardness. In addition to these main duties, a cutting fluid is often required to wash away the removed metal chips from the cutting area.

The fluids should contain no foaming agents or corrosive substances that might be harmful to the operator, the machine, the tool, or the workpiece. They must also have good 'wetting' properties so that they do not merely run off the workpiece immediately on contact.

There are three principal classes of cutting fluid in general use; light lubricating oils, emulsified oils, and chemical solutions.

25.3.1 Lubricating oils

Lubricating oils are used where cooling properties are a secondary consideration. An example is in using hand taps to produce a threaded hole. The cutting speed is low and the use of straight oil reduces the friction and protects the taps.

The three main types of cutting oil are: *mineral oils*, obtained as fractions during the cracking of crude oil and whose use is confined to light cutting operations; *fatty oils*, obtained either from animal or vegetable sources. The lubricating properties are good, making these oils the first choice for use with taps and dies; and *sulphurized oils*, sulphur-bearing mineral or fatty oils which operate extremely well under pressure conditions and reduce any tendency to welding between tool and removed metal chips. These oils are widely used for lathe work and other heavy-duty machining processes.

25.3.2 Emulsified oil

Emulsified oil is produced by rapidly whisking oil particles together with other substances designed to improve performance at the cutting face, in water. The main use of emulsified oils is as a coolant, the lubricating properties being of secondary consideration.

Emulsified oils being prepared for use in a machine should always be added to clean water. A breakdown of the suspension will occur if water is added to the oils. The mixing must always take place in a separate receptacle, never in the sump of the machine.

25.3.3 Chemical solutions

The chemical solutions used are oil-less substances. They consist of clean soft water containing detergents, rust inhibitors, and wetting agents. Their main purpose is as coolants for grinding processes.

Safety procedures must be observed when using cutting fluids because they often contain chemicals harmful to the skin. Provided that operators use the recognized brands of barrier cream on hands and forearms before work commences, and apply hand cleansing fluids or soap and plenty of hot water at the end of a working period, no danger should ensue.

Any cuts or grazes should be dressed with an antiseptic and a proprietory adhesive water-repellent bandage.

25.4 Summary

Cutting tools are basically wedge shaped and require rake and clearance angles to operate efficiently. The condition of the material from which the cutting tool is made should be such that it can be made extremely hard by heat treatment and will retain a keen edge under arduous conditions. To enable tools to function at maximum efficiency, cutting fluids are used to carry away the heat generated during machining and also assist in reducing friction between tool and workpiece. Cutting tools must be carefully sharpened and correctly set in relation to the workpiece in order to maintain the rake and clearance angles.

EXERCISES

1. Describe briefly the difference between a mixture, a compound, and an alloy.
   1.2, 1.4

2. Explain what is meant by the terms *mechanical properties* and *physical properties* of materials, and list three of each type of property.
   1.6

3. What is the difference between brittleness and toughness? Give an example of a brittle material and a tough material.
   1.6

4. Explain the difference between ductility and malleability, giving an example of a ductile material and a malleable material.
   1.6

5. Answer the following questions by adding the missing word or words.
   - (a) Materials capable of resisting very high temperature are called .... materials.
     1.5
   - (b) Carbon steel is basically an alloy of .... and
     .... 1.4, 3.1
   - (c) An elastic material is one which ....
     1.6
   - (d) Conductivity refers to the ability of a material to conduct .... and ....
     1.6
   - (e) Hard materials can resist .... and ....
     1.6
   - (f) Pig iron is melted in a .... to make cast iron.
     2.1
   - (g) Malleable cast irons are used to make electrical fittings such as ...., ..., ..., and
     .... Table 2.1
     2.2
   - (h) Silicon steel is used for the manufacture of ....
     3.5
   - (i) Annealing is a .... operation
     4.2.1
   - (j) A grain-refining process similar to annealing is called
     .... 4.2.2

6. Explain why some grades of copper can have a conductivity rating greater than 100 per cent.
   5.3

7. Why is brass not used in place of copper for conducting purposes?
   5.5.1

8. Explain the main difference between a soft magnet and a hard magnet. Why is sintering sometimes used to produce a magnet material?
   7.1, 7.2

9. Slightly magnetic materials are often used in temperature control devices. Why is this?
   7.4

10. Give five examples of electrical components that are made from non-magnetic, ferrous alloys and explain why non-ferrous metals could not be used instead.
    7.5

11. Name **three** metals, **two** gases and **one** non-metal commonly used as conductors of an electrical charge. Give the applications of the materials.
    8.3, 8.4, 8.7, 8.9, 8.11

12. Describe the desirable properties of electrical and telecommunications contact materials. Name **three** different contact materials and state their uses.
    8.5

13. Why is tungsten used as a filament in electric light bulbs?
Explain why it is necessary to evacuate the glass envelope   8.9, 8.10,
and name **two** gases sometimes used to replace the air.   8.11

14. Discuss the properties, uses, advantages, and disadvantages
of using aluminium as a conductor. Explain why steel reinforcing
is required in long-span aluminium overhead lines.   8.4

15. Answer the following questions by adding the missing word or
words.
    (a)   Tough pitch copper has had most of the ....
          removed.   8.3
    (b)   Cadmium copper is used for 8.3, 13.4
    (c)   Domestic three-pin plug contacts are made from
          ....   8.5
    (d)   Circuit breaker materials must have a high resistance
          to 8.6
    (e)   Carbon is used to make 8.7
    (f)   For low-arcing currents .... alloys can be used
          to make relay contacts.   8.8
    (g)   Discharge lamps use vapours from ....and
          .... to conduct an electrical charge.   8.11
    (h)   Filament support wires are made from 8.10
    (i)   The graphite in carbon brush material acts as a 8.7
    (j)   The strength of copper and copper alloy conductors is
          increased by 8.3

16. Describe the composition, uses, advantages, and disadvantages
of **two** copper base electrical resistance alloys.   9.2

17. Describe **one** method of producing bimetal strip and explain the
advantages of using this type of metal.   10, 10.1

18. Why are thermostat operating elements made from bimetals? What
are high-activity elements?   10.5

19. Discuss the advantages and disadvantages of using paper insulation.
Paper is hydroscopic; explain what this means and the methods
used to counteract it.   11.1

20. Enumerate **five** desirable properties of transformer insulating oil.   11.3.1

21. State the reasons for using insulating oil in switchgear.   11.3.1

22. Answer the following questions by adding the missing word or
words.
    (a)   Insulating wax is used to 11.3.2
    (b)   A thermoplastic material is one which 11.5
    (c)   P.V.C. is used for 11.5.1
    (d)   The disadvantages of using polythene as a sheathing for
          underground cables are 11.5.3
    (e)   Polystyrene cannot be extruded to form a cable covering
          because 11.5.5
    (f)   P.F. moulding powders are used to make 11.5.7
    (g)   P.F. laminated materials are used for 11.5.7
    (h)   Encapsulation of components is used to 11.5.10
    (i)   Mica insulating tape is made by 11.6.3
    (j)   Porcelain insulators are made from 11.8

Most
relevant section

23. Describe briefly, the method of making mineral-insulated copper
cable and state the precautions necessary when using this type
of cable.   **11.10**

24. Describe **two** conditions which can lead to insulation failure.   **11.12**

25. (i)   Explain briefly the way in which bimetal sheets are
          produced.   **10.1**
    (ii) State an application of a bimetal device and explain how
          it functions.   **10.3**
    (iii) During a sequence of switching operations in a direct   **10.4**
          current circuit a pair of contacts may fail due to 'welding'   **10.5**
          or 'sticking' together. State a reason for this.   **8.8**

                                                          C.G.L.I.
                                                            E.T.

26. Cast iron and aluminium alloys are two materials commonly used
    in general engineering. Discuss the merits and suitability of these
    materials in respect of:
    (i)   relative cost,   **Table 1.3, 2.1**
    (ii) the form in which the material is supplied to the primary   **6.3**
          processor,   **8.4, 14.1**
    (iii) subsequent processes,   **Table 2.1**
    (iv) typical applications.

                                                          C.G.L.I.
                                                            E.T.

27. Answer each of the following questions in one sentence only.
    (a)   Explain why 'killed spirits' should not be used for electrical
          joints.   **19.4.1**
    (b)   Explain the basic difference between an electrical conductor
          and insulator.   **8.1, 11.1**
    (c)   Why are joints in a transformer core kept to a minimum?   **12.1.1**
    (d)   Why is it not permissible to use brass as a former for a
          transformer winding?   **12.1.2**
    (e)   What is understood by the term *Volts per turn* applied
          to a transformer winding?   **12.1**
    (f)   What advantage has the H.R.C.[†] fuse over the rewireable
          type?
    (g)   Explain the reason for interleaving the layers of a
          transformer winding with insulating material.   **12.1.1**
    (h)   Why are armature conductors placed in slots in the
          armature core?   **12.1.1**
    (i)   Give a reason why an insulator may become conducting.   **11.12**
    (j)   Why are transformer coils and similar windings
          impregnated?   **12.1.2**

                                                          C.G.L.I.
                                                            E.T.

28. State the materials normally used in transformer construction
    for the following:
    (a)   the magnetic circuit   **12.1.1**
    (b)   the core insulation
    (c)   the conductor insulation   **12.1.3**
    (d)   the cooling medium.   **11.3.1**

    [†] An H.R.C. fuse is one with a High Rupture Capacity.

28. continued
     State the electrical properties of each of these materials and
     explain why they are suitable for the purpose stated.
                                                    C.G.L.I.
                                                      E.T.

29. Explain briefly, and with the aid of sketches, the following processes:
     (a)   copper wire drawing   17.1
     (b)   impact extrusion   17.3.1
     (c)   forging.   15.1
                                                    C.G.L.I.
                                                      T.T.

30. (a)   Explain why copper used for electrical work is generally of
           high purity.   8.3
     (b)   Explain briefly how copper can be annealed and what
           effect this has on its electrical conductivity. 5.3
     (c)   What reasons govern the choice between aluminium and   8.4
           copper for use as electrical conductors?   13.3
                                                    C.G.L.I.
                                                      T.T.

31. Name **two** insulating materials used in telecommunication cables.
     In regard to these, state   13.1
     (a)   with reasons, the type of cable each would be used for and   13.2
     (b)   how conductor identification is achieved in the design of
           each cable.   13.2
     For what ranges of cable size would **one** of the above insulating
     materials be used?
                                                    C.G.L.I.
                                                      T.T.

32. (a)   Explain the significance of the carbon content of steel and   21.3.1
           specify, with reasons, the type of carbon steel used for the   21.6
           making of small hand tools.   25.2
     (b)   Describe **three** of the major differences between iron and   21.5.1
           steel.   21.7
     (c)   Give an example in which steel would be a good choice of   Table 2.1, 2.1
           material for a workpiece but cast iron would not. Explain   Table 3.1
           why cast iron would *not* be a good choice in this instance.   3.4, 3.5
                                                    C.G.L.I.
                                                      T.T.

33. (a)   Comment briefly on the developments which have taken
           place in the materials used in cables for low-voltage
           distribution systems.   13.2
     (b)   List the advantages and disadvantages of tough rubber,
           polyvinyl chloride and mineral-insulated types of cable in   13.3.1
           (i)   extreme heat   13.4
           (ii)   extreme cold   11.4.1
           (iii) extreme humidity.   11.5.1
                                                    C.G.L.I.
                                                      E.T.

34.　Complete the following table.

Material	One typical application in electrical engineering	One reason for its use	
Aluminium foil			12.3.1 12.3.3
Aluminium			8.4
Copper			8.3
Brass			5.5.1
Silicon steel			12.1.1 16.2
Cotton			12.1.3 13.2
Magnesium oxide			13.3.1
Mica			11.6
Ceramic			11.8
P.V.C.			11.5.1

C.G.L.I.
T.T.

35.　(a)　Outline the causes of decay in wooden poles.　　　　　　　　13.5
　　　(b)　Describe the tests which should be made on a pole before
　　　　　climbing it.

C.G.L.I.
T.T.

36.　Answer all the following questions, which have been abstracted
　　　from past C.G.L.I. examination papers, by adding the missing
　　　word or words.

　　　(a)　A ductile material is able to withstand ....
　　　　　before fracture occurs.　　　　　　　　　　　　　　　　1.6
　　　(b)　The included angle to which a centre punch is ground is
　　　　　.......................... or ....　　　　　　　21.3.1
　　　(c)　A cutting compound acts as a ....　　　　25.3
　　　(d)　Hacksaw blades are fitted in the frame with the teeth
　　　　　pointing .... the handle.　　　　　　　Fig. 22.2
　　　(e)　A headstock is part of a ....　　　　　　24.2.1
　　　(f)　Nickel chromium wire is used to wind ....　12.2.1
　　　(g)　A .... is used to hold a die for thread cutting.　23.1.2
　　　(h)　.......................... is used as a flux for brazing.　　　　20.1.1
　　　(i)　Bimetal consists of two metals having different rates
　　　　　..........................　　　　　　　　　　　　　　　　10.5
　　　(j)　Transformer laminations are insulated from one another
　　　　　to ....　　　　　　　　　　　　　　12.1.1
　　　(k)　A reamer is a tool used to ....　　　　　22.1
　　　(l)　A drift is a tool used to ....　　　　　　Fig. 22.2

36. continued
   (m)   A pilot hole is used to **22.1**
   (n)   'Runner' and 'riser' are words used in the ....
         process.   **14.1.5**
   (o)   The degree of hardness of a metal is denoted by its

         ....   **1.6**
   (p)   A try-square is used for testing **21.2.3**
   (q)   Ceramic beads are used to insulate conductors operating

         in **11.9**
   (r)   Metals containing iron are called .... metals. **2.3, Fig. 1.1**
   (s)   Dowels are used to **14.1.3**
   (t)   Core prints are used in the moulding process to

         ....   **14.1.5**

37. State the materials used for the conductors, insulation and
     sheath of an underground cable. Explain the essential differences   **13.1**
     between twin, multiple-twin and star-quad type cables and what   **13.2**
     is meant by the lay of a cable.   **13.2**
                                             C.G.L.I.
                                             T.T.

38. Each of the following materials is either an insulator or a
     conductor of electricity.
     copper   aluminium   **5.3, 5.5.1, 6.3, 8.3**
     brass   p.v.c.   **8.4, 10.4, 11.4, 11.6**
     porcelain   rubber.   **11.8, 13.1, 13.2**
     (a)   Give an application for each of any five of the materials
           and two reasons for its choice in each application.
     (b)   Is an alternative available for the application you have
           chosen? If so, name an alternative material and state   **11.8**
           whether its use would be substantially different from that   **13.1**
           of the original material.   **13.2**
                                             C.G.L.I.
                                             T.T.

39. Discuss the characteristics desirable in a pole suitable for   **13.5**
     supporting overhead conductors.   **13.5**
                                             C.G.L.I.
                                             T.T.

40. Compare the accuracy of measurement of length obtainable using
     (a)   a steel rule   **21.2.1**
     (b)   a micrometer   **21.2.1**
     (c)   a vernier caliper.   **21.2.1**
     Assuming each of these measuring devices is capable of the
     accuracy demanded of it, explain for each condition how
     accurate measurement can best be obtained.
                                             C.G.L.I.
                                             T.T.

41. Comment on the properties desirable in the wire material for
     wire-wound resistors.   **12.2.1**
     Sketch and describe a typical wire-wound resistor for
     telecommunication application.   **Fig. 12.1**
     Can a more precise resistance value be obtained with this type of
     resistor than with a carbon type? Give brief comments.   **12.2.3**
                                             C.G.L.I.
                                             T.T.

42. (a)   List and compare the advantages and disadvantages
           of the following jointing processes
           (i)   soft soldering   19.5
           (ii)   brazing   20.1.1
     (b)   Comment briefly on the sources of heat available for
           these processes in (a) with particular reference to field   20.2
           work conditions.
                                                              C.G.L.I.
                                                                 T.T.

43. Describe, under the headings 'Object', 'Method' (or 'Procedure')
     and 'Conclusions', an investigation or experiment you have
     carried out on **one** of the following
     (a)   the use of various chisels for cutting metals   21.7
     (b)   the use of taps and dies   23.1
     (c)   the making of a permanent joint between metals.   20.2, 20.3
                                                              C.G.L.I.
                                                                 T.T.

44. (a)   Describe briefly the following processes
           (i)   sand casting   14.1
           (ii)   drop-forging   15.1
     (b)   Compare these processes in respect of
           (i)   strength of component produced
           (ii)   dimensional accuracy of component produced   15.1
           (iii) cost of component.
                                                              C.G.L.I.
                                                                 E.T.

45. (a)   Explain, with a sketch, the construction of a commercial
           can-type capacitor. State the materials used in its
           construction and give reasons for their choice.   12.3.1
     (b)   What are the basic differences in construction between
           paper dielectric and electrolytic capacitors?   12.3.3
     (c)   Explain the term 'working voltage', as applied to a
           capacitor.   12.3.3
                                                              C.G.L.I.
                                                                 E.T.

C.G.L.I. – City and Guilds of London Institute
E.T.   – Electrical Technicians
T.T.   – Telecommunications Technicians

INDEX

acetal resin, 73
acid iron, 13
afterblow, 14
alloying elements in
    aluminium, 37
    copper, 31
alloys, 4
alumel, 63
alumina, 34
aluminium, 34
    conductors, 45
angular measurement, 132
annealing, 21, 107
    full, 22
    spheroidize, 22
Araldite, 75
argon, 53

basic iron, 13
bauxite, 34
bench work, 127
bending, 103
beryllium copper, 45
Bessemer converter, 13, 14
bimetals, 59
blanking, 114
blast furnace, 9
blister copper, 28
blow, 13
blooms 17, 18
brass, 31
brazing alloys, 122
brightray alloys, 56, 57
brittleness, 6
bronze, 31
brushes, 50
bustle pipe, 10
butadiene-rubber, 69
butyl rubber, 68

cadmium copper, 45
calcine, 26
calipers, 127
cambric, 66
capacitors, 84, 85, 86, 87
carbon
    brushes, 50
    in cast iron, 10
    in steel, 19
carburizing, 23
case hardening, 23
case refining, 24
cast steel, 17

centre punch, 133
chaplets, 99
chemical cutting fluid, 158
chromel, 63
circuit breakers, 48
clearance angle, 152
cold
    -cathode tube, 53
    chisel, 139
    rolling, 105, 106
colour coding, 84
compound, 4
concentrate, 26
conductivity, 6, 43
conduit, 110
Constantan, 63
contacts, 47, 49
converter, 13, 27
coolant, 158
copper, 26, 29
    alloys, 31, 46, 55
    cladding, 60
    -nickel alloys, 55
core(s), 98
    prints, 99
    refining, 24
countersinking, 142
crucible steel, 17
cryolite, 34
cupola, 10, 11
cutting tools, 152, 156

die casting, 99
dielectric constants, 80
dies, 144
discharge lamps, 53
dolomite, 15
drawing, 102, 109, 114
drilling machine, 141
drills, 141
ductility, 6
duplex process, 16

ebonite, 68
elasticity, 6
elastomers, 67
electric furnace, 16
electrical contacts, 47, 49
    insulators, 65
    transmission lines, 89
electrolytic capacitors, 86
    refining, 28

elements, 3
    in cast iron, 10
    in copper alloys, 31
    in steel, 19
emulsified oil, 158
En steels, 19
epoxide resins, 75
Eureka, 63
eutectic solder, 116
evacuated envelopes, 52
extrusion, 109, 110

feeder head, 17
ferromagnetism, 38
ferrous metals, 12
Ferry, 56, 63
fettling, 98
fibreglass, 74
filament, 52
files, 136
filing, 136
fire refining, 28
flash welding, 124
flatness, 132
fluxes, 117
forging, 101
forming, 114
forms of supply, aluminium, 35
    cast iron, 11
    copper, 30
    steel, 18
full annealing, 22
fullers, 103
fusibility, 6

gas welding, 125
gases, 51
gas-filled envelopes, 53
glass fibre, 74
glass insulators, 78
graphite flakes, 11
grey cast iron, 11

hacksaw, 135
    teeth, 153
haematite ore, 13
hand scrapers, 138
hard magnets, 38
hardening, 22
    case, 23
hardness, 6
    of materials, 7
headstock, 147

heat treatment, 21
high
    -carbon steel, 18, 156
    -speed steel, 156
hot
    -cathode lamps, 53
    rolling, 105

impact extrusion, 110
impregnated paper, 65
ingots, 17, 18
insulating
    oil, 66, 67
    wax, 66, 67
insulation failure, 79
    of transformers, 82
insulators, 65
iron
    -base alloys, 56
    making, 9

lathe, 147
    headstock, 147
    saddle, 148
    steadies, 149
    tailstock, 147
    tools, 154
lead, 116
line supports, 90
linear measurements, 127
low
    -alloy magnets, 40
    carbon steel, 17
lubricating oil, 158

magnesium
    in cast iron, 12
    oxide, 78, 90
magnetic
    alloys, 40
    materials, 38
        permeability, 38
malleability, 6
malleable cast iron, 12
manganese
    in cast iron, 10
    in steel, 19
marking out, 132
matte, 27
measurement, 127
mechanical properties, 4
medium-carbon steel, 18
mercury vapour, 53
metal casting, 95
mica, 75
Micanite, 76
micrometer, 130
mild steel, 18
mineral insulation, 78, 90
mixture, 3

molecules, 4
Monel, 56
moulding flask, 96
    sand, 95

natural rubber, 68
negative temperature coefficient
    resistors, 55
neon, 53
Neoprene, 69
nickel-base alloys, 56
Nilo alloys, 61
nodular cast iron, 12
non-magnetic alloys, 41
normalizing, 22

odd-leg calipers, 135
off-hand grinder, 155, 157
open-hearth furnace, 15, 16
ore dressing, 26
oxygen-free copper, 30, 44

paper insulation, 65
patterns, 96
petroleum wax, 67
phenol formaldehyde, 73
phenolic laminates, 73
phosphor bronze, 31
phosphorus
    in cast iron, 10
    in steel, 19
physical properties, 4
piercing, 114
pig iron, 10
plasticity, 6
plastics materials, 70
platinum-group metals, 51
plumbing, 119
poling, 28
polyester resins, 74
polyethylene, 71
polystyrene, 71
polytetrafluoroethylene, 71
polyvinyl
    chloride, 70
    chloride acetate, 71
porcelain insulators, 77
power-cable insulators, 65
pressing, 114
pressure welding, 123
process annealing, 22
properties of
    aluminium, 34
    cast iron, 12
    materials, 4
protractor, 134

rake angle 152, 156
reamer, 142
refractory materials, 4

relay contacts, 50
resistance welding, 123
resistivity, standard of, 30
resistors, 82, 83
reverberatory furnace, 26, 27
rolling, 105
roundness, 132
rubber, 68

saddle, 148
sand moulds, 95
sandwich cladding, 59
scrapers, 138
screw cutting, 144
scriber, 133
scribing block, 134
seam welding, 124
semiconductors, 55
setting down, 103
shunts, 50
silicon
    bronze, 31
        in cast iron, 10
        in steel, 19
silicone rubber, 69
silver cladding, 60
    steel, 18
skip, 9
slabs, 17, 18
slag, 10, 13
sodium vapour, 53
soft magnets, 38, 39
solder, 116
    eutectic, 116
    silver, 122
    soft, 116
    tinman's, 117
soldering, 118
    irons, 119
solid solution, 4
spelter, 122
spheroidize annealing, 22
spot welding, 123
steelmaking, 13
stud welding, 124
standard of resistivity, 30
starting sheets, 28
sulphur
    in cast iron, 10
    in steel, 19
swages, 103

tailstock, 147
tapping, 144
taps, 154
teeming, 17
telephone cable insulators, 65
tellurium copper, 45
temperature coefficient of
    resistance, 55

tempering, 24
textile coverings, 79
thermo-electric series, 63
thermocouple materials, 63
thermostat materials, 61
three-jaw chuck, 149
tin, 116
tinman's solder, 117
tool geometry, 152
tough pitch copper, 28, 44
toughness, 6
tracking, 79
transformer(s), 8
    cores, 81
        insulation, 82

windings, 82
tungsten filament, 53
turning, 150
tuyère, 9, 10, 27
twist drill, 141, 153

underground cables, 88, 89
upset forging, 102
urea formaldehyde, 74

vacua, 51
vapours, 51
vernier caliper, 128
vinyl plastics, 70

vulcanized
    fibre, 77
    rubber insulation, 68

welding, 123
white cast iron, 11
wire
    bars, 30
    drawing, 109
wood poles, 90

xenon, 53

zinc, 29